Alfred Henry Fison

Recent Advances in Astronomy

Alfred Henry Fison

Recent Advances in Astronomy

ISBN/EAN: 9783337395872

Printed in Europe, USA, Canada, Australia, Japan

Cover: Foto ©berggeist007 / pixelio.de

More available books at **www.hansebooks.com**

The Victorian Era Series.

In crown 8vo volumes, cloth, 2s. 6d. each.

The series is designed to form a record of the great movements and developments of the age, in politics, economics, religion, industry, literature, science, and art, and of the life work of its typical and influential men.

Under the general editorship of Mr. J. HOLLAND ROSE, M.A., late Scholar of Christ's College, Cambridge, the individual volumes are contributed by leading specialists in the various branches of knowledge which fall to be treated in the series.

The Rise of Democracy.
 By J. HOLLAND ROSE, M.A.

The Anglican Revival.
 By J. H. OVERTON, D.D., Canon of Lincoln.

John Bright.
 By C. A. VINCE, M.A., late Fellow of Christ's College, Cambridge.

Charles Dickens. By GEORGE GISSING.

The Growth and Administration of the British Colonies, 1837–1897.
 By the Rev. W. P. GRESWELL, M.A., author of "Africa South of the Zambesi", "History of the Dominion of Canada".

The Free-trade Movement and its Results.
> By G. ARMITAGE-SMITH, M.A., Principal of the Birkbeck Institution, and for many years Lecturer on Economics for the London Society for the Extension of University Teaching.

English National Education.
> By H. HOLMAN, M.A., formerly Professor of Education in the University College of Wales, Aberystwyth.

Provident Societies and Industrial Welfare.
> By E. W. BRABROOK, C.B., Chief Registrar of Friendly Societies.

London in the Reign of Queen Victoria, 1837-1897.
> By G. LAURENCE GOMME, F.S.A.

Recent Advances in Astronomy.
> By A. H. FISON, D.Sc.(Lond.).

Forthcoming volumes, published Monthly.

Charles Kingsley and the CHRISTIAN SOCIAL MOVEMENT.
> By the Very Rev. C. W. STUBBS, D.D., Dean of Ely.

The Science of Life. By J. ARTHUR THOMSON, M.A.

Indian Life and Thought since the Mutiny.
> By R. P. KARKARIA, B.A., Principal of the New Collegiate Institution, Bombay, and Examiner in History and Philosophy to the University of Bombay.

Alfred, Lord Tennyson: A CRITICAL STUDY.
> By STEPHEN GWYNN.

British Foreign Missions.
> By Rev. WARDLAW THOMPSON and Rev. A. N. JOHNSON, M.A.

⁎⁎⁎ Prospectus and press opinions may be had on application.

LONDON: BLACKIE & SON, LIMITED, 50 OLD BAILEY, E.C.
GLASGOW, AND DUBLIN.

The Victorian Era Series

Recent Advances in Astronomy

Recent Advances
in
Astronomy

By

ALFRED H. FISON, D.Sc.

LONDON
BLACKIE & SON, Limited, 50 OLD BAILEY, E.C.
GLASGOW AND DUBLIN
1898

Preface

In the following pages I have endeavoured to give a simple account of some of the more interesting "Recent Advances in Astronomy". To harmonize with the general scheme of the series of which this work forms a volume, it was at first suggested that I should develop recent progress in Astronomy historically. The difficulties in the way of treating any branch of science in such a manner are, however, very considerable; especially when, as in the present instance, it is desired to present the subject in such a manner as to be readily followed by those who have but slight familiarity with its technicalities. I am only acquainted with one entirely satisfactory "History of Astronomy", and that one scarcely appeals to other than professional astronomers. It has upon the whole appeared best to effect a compromise between an historical and a purely descriptive method; and I have, therefore, while dealing with what have appeared to me to be a few among the more interesting problems of modern Astronomy in a series of separate essays, followed in each the historical method as far as possible. It has been found practicable to adhere to this scheme more rigidly in the latter part of the work.

Every writer of a popular work on Astronomy, or any other branch of science, must become largely indebted to those who have devoted their labour to the compilation of works of reference; and I would acknowledge my deep obligation to the extensive accumulation of accurate knowledge contained in Miss Clerke's two works—*A History of Astronomy during the Nineteenth Century*, and *The System of the Stars*.

A. H. FISON.

September, 1898.

Contents

CHAPTER I
The Life of a Star - - - - - - - - 1 *(Page)*

APPENDIX TO CHAPTER I
The Measurement of Stellar Distances - - - 50

CHAPTER II
The Milky Way and the Distribution of Stars - - 59

CHAPTER III
The Recent Study of Mars - - - - - - 101

CHAPTER IV
The Analysis of Sunlight - - - - - - 144

CHAPTER V
The Analysis of Starlight - - - - - - 193

CHAPTER VI
The Red Flames of the Sun - - - - - - 219

INDEX - - - - - - - - 239

Recent Advances in Astronomy.

Chapter I.

The Life of a Star.

> "Great is the mystery of Space, greater is the mystery of Time. Either mystery grows upon man, as man himself grows; and either seems to be a function of the godlike which is in man. In reality, the depths and the heights which are in man, the depths by which he searches, the heights by which he aspires, are but projected and made objective externally in the three dimensions of space which are outside of him."
>
> De Quincey.

With our present knowledge of the sun-like nature of the stars, and the colossal part that they play in the scheme of the physical universe, it appears strange that, in spite of the bold spirit of speculation that characterized the ancient philosophy,—a philosophy that recognized the possibility of the development of higher forms of life from lower; that saw in the Sun, Moon, and Earth different forms of air in different stages of condensation; and in the universe itself the working of a fortuitous concourse of atoms,—no worthy speculation should have been recorded as to the nature of the stars.

Alike to the philosophers of Ancient Greece, and to the early astronomers of Greece and Alexandria whose lives were spent in tracing with deepest thought and rarest skill the movements of the heavenly bodies, it was sufficient that the stars were points of fire, each set in its place in the concave of the firmament, and eternally borne by it in diurnal revolution round the central Earth.

It is unnecessary to do more than very briefly review the steps, initiated in the bold speculations of Copernicus towards the middle of the sixteenth century, by which our present knowledge of the sun-like nature of the stars has been attained. Copernicus had shown it to be probable that the Earth is one of the planets, a group of small bodies revolving, each in its own period or year, round the central Sun; and had recognized, as the logical consequence of his scheme, that to remain, as they appeared to remain, unaffected in their apparent positions upon the celestial vault during the supposed annual sweep of the Earth in its orbit, the stars must be vastly more remote than the Sun; but to him the material vault of heaven had merely been thrown farther back, and the stars were still points of fire studding its concave surface. The bolder and direct deduction—that to appear unaffected in direction through all seasons of the year, in spite of the enormous displacement in the position of the observer upon the moving Earth, the stars must be so remote, that, to be visible at all, the majority of them must be bodies of the same order of light-giving power with the Sun itself—was recognized, though with hesitation, by John Kepler, and was for the first time fully accepted by Galileo.

It will be advisable to present the principle underlying this deduction in a more definite form, since the thorough comprehension of what has already been achieved by it, and what may reasonably be expected from its application in the future, is of fundamental importance in the astronomy of the stars. It involves directly the only method that has so far been successfully applied to the measurement of the distance of a star.

Let the curve in fig. 1 be regarded as representing the orbit of the Earth round the Sun, an oblique view of the nearly circular orbit, and suppose that, the Earth being in the position indicated by the point P, a star is observed, and that the direction in which it is seen is recorded with all possible accuracy. Let the straight line PA indicate this direction. The star must lie somewhere in the direction PA, but there is nothing in the observation to indicate its distance from A, the point of observation. Six months later, however, the Earth will have reached the position indicated by Q, having traversed in that time one-half of its complete orbit.

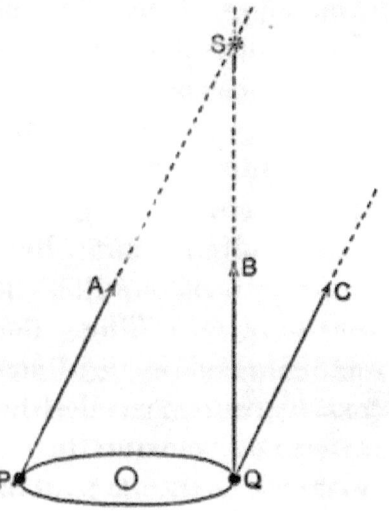

Fig. 1.—Illustrating Stellar Parallax.

Let the direction in which the star now appears be observed, and let it be recorded by the line QB. The star lies therefore in the direction QB, but there is nothing in the last observation to indicate its exact position in this direction. Since, however, the pair of observations have located the star in the directions PA and QB, it must be situated at S, their point of intersection, and the geometry of the figure at once gives the proportion between the distance of the star and the dimensions of the Earth's orbit.

The principle of this, the only method by which the distance of a star has so far been determined, cannot but appear extremely simple, but a difficulty in the interpretation of its application appeared, when to Copernicus, as to his followers for nearly three hundred years, the lines of sight PA and QB appeared to be parallel, showing no tendency whatever to meet. Thus, the first observation having established PA as the direction of the star, the second would give the parallel line QC, and not a direction such as QB, sensibly inclined to the first.

Assuming the fact of the Earth's journey round the Sun, the only possible interpretation of the apparent parallelism of the lines of sight was that the stars are so remote that, although the directions PA and QC are inclined toward each other, each being directed to the star, the inclination is so slight that it was incapable of detection. That this was the true explanation appeared more and more certain as the truth of the Copernican system became more firmly established; and in the conviction that success was possible, and as instruments were devised by the employment of which it became possible to determine with an ever-increasing degree of accuracy the direction of a star, the search after the inclination of the lines of sight towards a star from opposite extremities of the Earth's orbit, or the "parallax of a star", became increasingly keen.[1]

It is a matter of history that after close upon three centuries of arduous toil—toil occasionally rewarded by unexpected discoveries of the greatest interest and of the farthest-reaching importance, though resulting in failure so far as the immediate object of the search was concerned—success was at last achieved. In 1838, Bessel of Königsberg demonstrated, as the result of the critical examination of a great number of observations, that a certain small star in the constellation of the Swan did appear to experience a displacement in its position upon the heavens during the progress of the year, the inclination of the lines of sight toward it from the two most favourably situated positions of the Earth in its orbit being estimated at nearly two-thirds of a second of arc. The star, in itself an insignificant member of the orbs of heaven, thus destined from its association with Bessel's discovery to acquire an honourable place in the history of astronomy, is known as 61 Cygni; and its distance as deduced from Bessel's measurements was 600,000 times that of the Sun. Bessel's estimate has, however, been reduced by

[1] The parallax of a star is more exactly defined in astronomy as one-half of the greatest observable inclination of the lines of sight, or as the inclination toward each other of two straight lines to the star, one directed from the Sun and the other from the Earth at a time when the direction of the Earth, as seen from the Sun, makes a right angle with that of the star.

more recent measurements, carried out with finer instrumental appliances, and with the advantages arising from accumulated experience, to 440,000 times the distance of the Sun.

It would be scarcely advisable to digress here into a critical examination of the difficulties that have been experienced in the search after stellar parallax, and the methods by which they have been in part overcome; but the subject is of such high importance in the astronomy of the stars that I have ventured to give, in the form of an appendix to the present chapter, a rather more detailed account of Bessel's discovery, as well as of the leading features of more recent work.

Within a few months of the date of Bessel's discovery, Professor Henderson, of Edinburgh, announced the fact of his having succeeded in detecting parallax in the bright southern star α Centauri. The observations in which the parallax of the star was recorded had been made by Henderson six years previously, and for a different purpose, during the course of his work at the Cape of Good Hope, and his attention was only redirected to them with a view to the investigation of parallax by the announcement of Bessel's success. The distance of α Centauri, deduced from the parallax originally announced by Henderson, is 180,000 times that of the Sun, but the most recent measurements have extended it to 270,000 times the distance of the Sun.

So far, no star has been found to lie nearer to the solar system than α Centauri. A vague suggestion of the unthinkable void that separates us from this, in all probability the nearest of the stars, may perhaps be obtained from the fact, that upon such a scale that the orbit of the earth should be represented by the circumference of a shilling, the star would be removed to a distance of two miles. Across such a distance as that of α Centauri, light, travelling with a velocity of 187,000 miles in a second of time, would speed onward for four and a half years; so that the star is seen, not as it is now, but as it was four and a half years since, while another equal period must pass before the rays now leaving it will bring their record to the shores of the Earth.

The discoveries of Bessel and Henderson imparted new life to the search after the parallaxes of stars. So delicate, however, are the necessary observations of direction, and so many and serious are the sources of error, that, excepting a few isolated successes, the record of the next forty years is chiefly one of accumulation of experience; and when in 1881 Dr. Gill and Dr. Elkin commenced a series of observations at the Cape of Good Hope, the parallaxes of not more than half a dozen stars had been detected with certainty. Since that date, however, parallax hunters have been better rewarded, though up to the present time it is doubtful whether success has been achieved in more than fifty instances.

Of the stars the parallaxes of which have been detected, Sirius is undoubtedly the most important. As in the case of α Centauri, its parallax is indicated with some degree of probability in observations made by Henderson in 1832. For another half-century, however, the numerous attacks made upon it were chiefly remarkable for the discordance of their results, discordance that ultimately vanished in the frequently repeated observations that have been made at the Cape since 1891. The most recent estimate—published by Gill in 1898—of a parallax of ·37 of a second of arc, agrees very closely with his previous results, and indicates for the star a distance of 556,000 times that of the Sun.

The mere statement of the distances of stars is apt to be productive of weariness of the spirit; in their absolute magnitudes they are entirely unthinkable, but in their relation to things familiar, they may well produce a powerful impression of the nothingness of the Earth—so far as its physical relations are concerned—to the scheme of the physical universe. From the days of the Psalmist it has been customary to regard the heavens as inspiring a sense of deep humility in man, but whether Nature in her most sublime aspect would appeal to one who had not already learnt the lesson from communion with his fellows is doubtful.

The chief interest of the distances of the stars in the present connection lies in the view to which they necessarily lead regarding the nature of the stars themselves. If it were removed to the distance of Sirius, the Sun itself would fade into insignificance, shining but as a star of the third magnitude, rather less conspicuously than the brighter ones that form the familiar W of Cassiopeia. Seventy-five such stars would be necessary to supply light equal to that received from Sirius; hence, in the intensity of its light radiations, Sirius exceeds the Sun 75 times. Of the few stars in which parallaxes have so far been detected, to appear with their actual luminosities at their estimated distances, many must far exceed the Sun in light-giving power, while a few must surpass even Sirius itself. Others, however, and among them Bessel's star in the Swan, while undoubtedly suns, would appear but as modest specimens of their class if placed beside ours, and it is scarcely possible so far to decide whether our Sun reaches the average of splendour displayed by the suns of space, or whether he surpasses it. It will be seen in a later chapter that the sun-like nature of the stars is further indicated in the analysis of their light by the spectroscope.

When, as is the case in the overwhelming majority of instances, no parallax can be detected in a star, its distance is of course indeterminate, but it is possible to assign a minimum distance beyond which it must be situated, if the smallest angle of parallax that could escape detection is known. So much depends upon the skill of the observer, upon the position of the star in relation to others, and even upon its colour, that it is not possible to give any definite and general estimate of this maximum parallax. According to a recent statement of Dr. Gill, however, than whom undoubtedly there can be no higher authority, under favourable conditions a parallax of a fiftieth of a second of arc should not escape detection, one that corresponds to a distance of rather more than 10,000,000 of times that of the

Sun. There is little doubt that but an insignificant fraction of the stellar host lie within this limit.

To apostrophize upon the picture of the physical universe revealed by these discoveries is an old story. The concave vault of the Old Astronomy has dissolved, and has revealed, beyond, a scheme unthinkable in its vastness, and in its suns and systems of suns radiant with energy. There are few among us who have not experienced, as in our more emotional moments we have endeavoured to penetrate, however superficially, the inward mystery of so majestic a scheme, and one in which man plays apparently so humble a part, a sense of oppression. We have been overwhelmed with the sense of inscrutable and immanent mystery; and we have been ready to exclaim with the pilgrim of German fable, "I will go no farther; for the spirit of man acheth with this infinity. Insufferable is the glory of God! Let me lie down in the grave and hide me from the persecution of the Infinite, for end, I see, there is none!"

The stars, then, are suns; and the life of a star is the life of a sun. Life is essentially a succession of changes, a passage through varying conditions of activity; death is cessation of all activity. Are there grounds for regarding our sun as undergoing change? and if there are, what is the nature of that change? Are there indications that in time the activity of the Sun will cease? The Sun is clearly a hot body continually throwing off an enormous amount of heat into space by the process of radiation. Unless, therefore, by some undiscovered and entirely unsuspected process an equivalent amount is supplied to it from some external source, it must be becoming continually poorer in its store of heat.[1] In the remote past it must have contained far more, and in the distant future it will contain far less heat than it contains at the present time. Everything goes to support the straightforward view that the light of the Sun is the direct result of a vivid state of incandescence of its surface consequent upon the high temperature to which it is raised. As the Sun cools, a time must come when, unless some catastrophe intervenes, its temperature will have fallen so much that its beams will have lost their present glory; later, it will cease to glow; and thereafter, as a dark star, a sad memorial of its present splendour, it will pursue its lifeless course through the ages.

Attempts have been made to estimate the time that must elapse before the period of this Sun-death, but in our ignorance of the physical constitution of the Sun, and more especially of that of its interior, such estimates are affected by a very wide margin of uncertainty. There is, however, no doubt that the actual heat existing as such in the Sun forms but an insignificant fraction of its total store of radiation. Without doubt, the Sun is largely, if not almost wholly, gaseous: and since all gases, as also nearly all solids and liquids, expand with

[1] A rather attractive speculation of Julius Mayer, vigorously supported for a time by Tyndall, sought to account for the maintenance of the solar radiation by heat developed from the destruction of motion of meteorites continually falling into the Sun. It has, however, been shown that any heat that the Sun may possibly gain in this way must be quite negligible in quantity.

accession of heat and contract upon loss of it, the sun must be shrinking. Further, the interior of the Sun must be enormously compressed by the weight of superincumbent matter; and as shrinkage takes place it must become still more compressed. But the act of compression of a gas produces heat; hence heat is continually being generated in the Sun by the compression of its substance. Each step in the loss of heat, therefore, calls into existence other heat; and this may partly, wholly, or even for a time more than compensate, for the loss. Subjecting these principles to mathematical expression, Helmholtz has shown that the heat thus evolved by compression of the Sun consequent upon its shrinkage must be sufficient in amount to maintain it as a self-luminous body for many millions of years. It is unnecessary for our present purpose to attempt to arrive at a more definite estimate of the future of the life of the Sun.

Returning to the present condition of its system, and from it projecting our thought backward into the past, we see the Sun richer and richer in heat during receding ages; becoming more perfectly gaseous—the ultimate effect of accession of heat being to convert all things into the gaseous state—while ever increasing in volume; the planets one by one disappear in its expanding bulk; and there appears as the first stage of Sun-life a diffused body of gas, extending beyond the present limits of the planetary system, and containing latent in itself a store of energy that is through coming ages to maintain the vitality of worlds.

The Sun, and, by the same process of reasoning, the stars, would thus appear to have originated in extended volumes of tenuous gas, and to be fated in the end to be degraded into cold inert masses. These conclusions being accepted, it would appear probable that both of these conditions would be at the present time represented among celestial bodies, for even upon the extreme assumption that all of them were created at the same time and in the same stage of development, it would follow that, since they differ enormously in mass, they would cool and therefore pass through their life stages at different rates. It becomes, therefore, of great interest to inquire whether there exist in celestial space extensive bodies of gas, and whether there exist dark stars. The answer is clear: astronomical observation has revealed both.

There can be little doubt that the earliest stage of star-life is represented in, at any rate many of, the nebulæ. The nebulæ appear as faint clouds of light, and are distributed in thousands over the face of the heavens. The greater number are excessively faint, their very detection demanding the aid of the highest optical power; while two only, and those just hovering upon the verge of vision, are visible to the eye upon the darkest and clearest nights. These are the glorious objects in the constellations of Andromeda and Orion, the one in Orion being the more impressive of the two.

The Great Nebula of Orion is situated near the centre of a line of faint stars that trail southward from the middle of a line formed by the three bright ones that constitute the "belt" of the familiar winter constellation Orion. Visible to the naked eye under favourable conditions as a faint mist—

> A single misty star,
> Which is the second in a line of stars
> That form a sword beneath a belt of three;—

its cloudy nature clearly revealed in a hand telescope or a good field-glass; when viewed through a telescope of large light-grasping power it becomes one of the most impressive of natural objects, though the vast extension of the heavens into which its wreaths are thrown, and the abundance of delicate detail permeating the whole, have only become revealed in recent records of the photographic plate.

The Nebula of Orion appears through a fine telescope as a faint green haze, suggesting a light cloud floating in celestial space, in form not very unlike that of the profile of a fish's mouth. The whole is composed of clouds of light of different degrees of brightness, some of extreme fantastic, and not a few of highly suggestive forms. It is in the perception of these that the photographic plate has demonstrated, as powerfully as in any of its applications, its great superiority over the eye in its capacity of appreciating the faintest shades of light. Structure is revealed throughout the whole nebula by the manner in which streams of luminous matter are directed from a brilliant and nearly central region in close proximity to the mouth-like bay of dark sky, the importance of this region being emphasized by the occurrence in it of a remarkable group of stars—" the trapezium of Orion "—and in the symmetrical arrangement of many of the cloud-forms with reference to it. Many stars are scattered over the picture; that those of the trapezium are actually involved in the glowing wreaths of the nebula itself, and do not owe their appearance in it to the effect of optical projection, either by their lying by chance in the line of sight towards the nebula, or by being visible through its transparent substance while actually far beyond, is rendered overwhelmingly probable from their position with reference to the cloud-forms, as well as by certain relations that have been shown to exist between their analysed light and that of the immediately surrounding nebula, in the spectroscopic researches of Sir William Huggins.

The diffuse character of the outlines of the nebula renders it impossible to apply to it such delicate measurements of direction as are necessary for the determination of the parallax. For this reason its distance cannot be directly investigated. The stars of the trapezium have, however, shown no parallax; from this it becomes possible to assign roughly a minimum limit beyond which they, and therefore in all probability the nebula, must lie. Such distance can scarcely be less than a million times that of the Sun. To appear of its vast extent, even at this, the most modest estimate, its glowing clouds must extend over such abysmal depths, that the whole of the Solar System if plunged into it would become contemptible in its utter insignificance.

The Nebula of Orion is a noble example of an "irregular nebula". That of Andromeda, in its regular ellipticity of outline, in the uniformity in the central condensation of its light, and in the system of elliptical rings by which it is enveloped, forms so strong a contrast with it that it is difficult to regard the two as objects belonging to the same class. Other nebulæ display a spiral structure; others again appear as fairly sharply defined planetary discs; while the majority are to all appearance nothing more than minute structureless clouds of flocculent light.

In the early period of their discovery, a discovery that followed naturally upon Galileo's first application of the telescope to astronomical observation in 1609, nebulæ were regarded as diffusions of a lucid medium shining by its own inherent lustre. In 1780, the year that marked the commencement of Sir William Herschel's classical researches upon them, less than 150 were known; but as the result of those researches, which extended over a period of twenty-one years, their number had been increased to close upon 2500. By their extended distribution in space, as well as by the detailed structure revealed in many of them by Herschel's observations, the nebulæ had acquired a new importance in the system of the Universe.

From a not altogether satisfactory deduction from the universality of gravitation, an extension of natural law that his own discovery of the mutual revolution of the components of double stars went far to establish, Herschel was led, in the earlier period of his researches, to reject the generally received view regarding the nature of nebulæ, and to substitute for it one according to which they were clusters of stars, the component stars being too faint, by reason, it was supposed, of excessive distance, for their individuality to be recognized. While maintaining this view with regard to the constitution of some nebulæ, Herschel, however, subsequently reverted to the former hypothesis to account for many of them, these including the Nebula of Orion, regarding them as "extensions of a shining fluid of a nature unknown to us". He further framed a first consistent scheme of stellar evolution, in suggesting that individual stars and clusters of stars were formed by the condensation of this nebula substance by the power of gravitation.

During the first half of the present century scientific opinion entirely reverted to the earlier of Herschel's views. Changes in the outlines of certain nebulæ, and the absence of structure in others of the "planetary" class, both of which Herschel, thinking that he had established by observation, had advanced in support of his later views, failed to receive confirmation in their examination by later astronomers. As with increased telescopic power many objects classed as nebulæ were one by one resolved into collections of stars, the conviction became increasingly strong, that, with sufficiently refined means, all would ultimately succumb: and when at length, in 1850, the Great Nebula of Orion was thought, from its appearance in the gigantic telescope of Lord Rosse, to show indications of breaking into clouds of stars, the riddle of the nebulæ appeared to be approaching its last solution. As clusters of stars the nebulæ found ready place in the speculations of many astronomers, whose minds, in consequence of the perfection displayed in the relations between the Sun and planets, had become powerfully impressed with the conception of a system as an essential unit in the construction of the universe. The planets, with their attendant satellites, formed systems, fair images of the grander Solar System, in which they were included. Each star was regarded as a sun, the centre of a planetary system of its own. Visible isolated stars formed with our Sun a larger but essentially similar system or "galaxy", in which it was even conjectured that all members might revolve round a central orb; while nebulæ were other systems of suns, external galaxies, awfully remote from our galaxy and from each other; oases of active energy scattered through space. The demolition of this stupendous conception by later researches has been advanced as supplying the only instance in which astronomical discovery has failed to reveal in the actual a more majestic scheme than had previously been idealized in the boldest imagination.

While, however, the colossal reflector of the Earl of Rosse was engaged, it was fondly believed, in finally establishing the nebulæ as clusters of faint stars, the researches of Ångstrom, Bunsen, Kirchhoff, and others were placing upon a firm foundation the principles of a new science that was shortly to enter the arena, with the result of utterly confounding general expectation. The development of the science of Spectrum Analysis forms the subject of a later chapter of the present work. Here it must be sufficient to record that as early as 1672 Sir Isaac Newton had shown that, upon passing a ray of sunlight through a glass prism, it became separated into its constituent colours, by reason of the fact that all rays are deflected or "refracted" on traversing the prism, but that rays of different colours are refracted to different degrees; that after the lapse of a century and a half the study of the analysis of light was resumed and the instrumental means greatly improved by Fraunhofer of Munich; and that, by the labours of Kirchhoff and Bunsen, the spectroscope assumed its place as a powerful instrument of research about the year 1860.

The spectroscope is essentially an instrument whereby light consisting of a mixture of colours is, after entering the instrument by a narrow slit, resolved into its constituent colours by a prism, or occasionally by an equivalent "diffraction grating". The separated colours are in either case spread out into a tinted band or "spectrum". About the middle of the present century observations with the spectroscope had indicated that there was a remarkable difference between light emitted by a glowing gas and that radiated from an incandescent solid or liquid body. With light emanating from an incandescent solid or liquid, such as that emitted by a glowing mass of white-hot metal, or by a gas flame in which the greater part of the luminosity is due to incandescent clouds of soot deposited in the flame from the decomposition of the gas under the intense heat of combustion, and, with a limitation to be noticed subsequently, that from the Sun and from the great majority of the stars, the spectrum consists of a continuous band in which all the colours of the rainbow are represented, each passing into the next by insensible gradations, while red and violet occupy the extreme positions. In the light from a glowing gas, however, at any rate when the density of the gas is not excessive, this is not the case, the light being now resolved into a series of clearly-defined and separate colours, which appear in the spectroscope as bright lines of coloured light separated by dark intervals; the lines are, in fact, images of the slit by which the light enters the instrument, a separate image being formed by each of the colours present. The light from the flame of a spirit-lamp which has acquired a strong yellow tint by sprinkling a trace of common salt upon the wick, is, for instance, resolved into two closely coincident shades of yellow, indicated in the spectroscope by the appearance of a pair of closely adjacent yellow lines; and the peach-coloured glow emitted by hydrogen gas when rendered luminous by a discharge of electricity through it, gives rises to the appearance of several coloured lines, of which a crimson and an emerald-green appeal most strongly to the eye.

In the year 1864 Sir William Huggins first

applied the spectroscope to the study of the nebulæ, the particular one selected being a small but comparatively bright object in the constellation of the Dragon. The light from the nebula was condensed upon the slit of the spectroscope by the object-glass, 8 inches in diameter, of an astronomical telescope; and at the first glance, the examination of the spectrum showed it to be characteristic of the light emitted from a glowing gas, since it consisted, not of a continuous band, but of three separated lines, all of them being of a green colour. The luminous matter of the nebula consisted, therefore, not of a host of stars, but of incandescent gas; and the more matured views of Sir William Herschel were established upon a sound scientific basis.

During the four years following this observation Huggins subjected the light from seventy other nebulæ to analysis; and of them about one-third, including the Great Nebula in Orion, proved to be gaseous. The remaining two-thirds yielded "continuous" spectra, spectra in which all shades of colour were represented, and might, therefore, so far as spectroscopic evidence was concerned, consist of systems of stars, of gas possessing comparatively high density, or of gas in an incipient stage of condensation. The structure of some of these as revealed by the photographic plate lends strong support to the last hypothesis; in the Great Nebula in Andromeda, for instance, it is scarcely possible not to recognize the process of condensation as actually in progress. Nearly one-half of the nebulæ owe their luminosity to the presence in them of glowing gas.

It is difficult not to see in the gaseous nebulæ the stuff of which future stars will be made. Granting that their substance is subject to the law of gravitation, it appears certain that in coming ages their glowing matter must, under its influence, be drawn towards centres of condensation; the smaller and more symmetrical of the nebulæ possibly developing into single stars, but such majestic collections of cloudy structures as are revealed in Orion being more probably the origin of hosts of separate suns.

Turning from these impressive representations of the birth of suns, it now becomes our task to seek among the heavenly bodies for the more sombre but scarcely less impressive record of their death; to search among their resplendent brethren for evidence of the existence of spent and dark suns. A dark star may conceivably become known to us in either of two ways: it may in its wanderings through space interpose itself between the Earth and a bright star, thus producing a total or a partial eclipse of the latter; or it may approach sufficiently near a visible star to affect it sensibly by its gravitational influence, in which case it may be possible to deduce the existence of the dark star from the disturbance apparent in the movement of the bright one. There can be no doubt that the existence of dark stars has been revealed in both of these ways, and both methods of research are admirably illustrated in the discovery of the notorious dark companion of Algol.

From the extremity of the shallower and left arm of the familiar W of Cassiopeia, and setting off in a direction making sensibly a right angle with the limb, a gracefully curved line is naturally traced in the heavens by the stars of Perseus and terminated in the resplendent orb of Capella. A straight line diverging to the left of this stream and proceeding slightly forwards from the brightest and nearly central star in Perseus, is directed to Algol, the best-known of the variable stars.

The variable character of the light of Algol is said to have been first observed by Montanari in 1669, though, owing to the great difficulty in measuring the intensity of starlight, it has only been possible in recent years to trace the exact law of its variation with any approach to scientific accuracy. Normally, Algol appears as one of the conspicuously brilliant stars of the heavens, its brightness being sensibly the same as that of the Pole Star. At intervals of time that appear to be subject to a very slow variation, and which are at present represented by 2 days 10 hours 48 minutes and 52 seconds, its light commences to fade, and continues to do so for 4½ hours, by which time it has decreased to two-fifths of its normal brightness. This minimum value it retains for 20 minutes, after which it resumes its normal lustre in a manner which is nearly, though not exactly, the reversed image of its fading.

In 1782 Goodricke, impressed with the regularity displayed in the repeated variations of the star's light, suggested as the cause of it the existence of a dark companion revolving round Algol in an orbit presented edgeways to the Earth, so that at each revolution the bright star would suffer partial eclipse by the interposition of its companion between it and the Earth. The explanation was obviously sufficient to account for the mere fact of periodic variation, and its truth appeared more probable, when, a century later, the spectroscope showed the variation of the star's light to be unaccompanied by any change in its quality. Such change would indicate change in the star itself, and is frequently a conspicuous feature in the variation of other and less regularly variable stars. The probability of the truth of the eclipse theory of Algol was still further increased when in 1888 Professor E. C. Pickering of Harvard, by the application of the "meridian photometer", an instrument by the invention of which it became possible to measure the intensity of the light of a star with a degree of accuracy previously unattainable, found, from the examination of the light of Algol at repeated intervals during the progress of its variation, the law or method of its variation to be essentially such as would result from the interposition of a dark sphere between the Earth and a luminous one.

Closely following upon Pickering's researches, and by the application of a principle suggested by him, the final demonstration of the existence of Algol's dark companion was effected by Vogel at Potsdam, from observations made between the years 1888 and 1891. Assuming the existence of a star revolving round Algol, it would appear probable that the force necessary to constrain it to continually follow its curved path would be found in gravitational attraction exercised upon it by Algol. By such a force, the attraction of the Earth, the Moon is maintained in its nearly circular path around it; by such forces, the attractions exercised upon them by the Sun, the planets follow without deviation their determined orbits. If, however, Algol attracts its companion, it follows from the necessary equality between action and reaction as expressed in Newton's Third Law of Motion, that the companion must attract Algol with an equal and opposite force, and it is conceivable that motion of Algol caused by the attraction of the companion might be capable of detection.

The character of the motion of two mutually attracting bodies was first determined by Newton. It was shown by him to follow from the laws of motion, combined with the fact that the attraction of gravitation varies inversely with the square of the distance separating the attracting masses, that the pair must describe similar conic sections having a common focus, which is continually occupied by the centre of mass of the pair.[1] Which of the three possible forms of conic section will be assumed by the orbits depends upon the initial circumstances of the motion, but the greatest interest is attached to the ellipse, which, being the only conic section forming a closed curve, must be the orbit in every case in which the motion is repeated.

The mutual revolution of the Earth and Moon supplies an interesting illustration of the nature of the motion under consideration. The Moon is

[1] The term "centre of mass" corresponds to the point more generally known in elementary mechanics as "centre of gravity". For obvious reasons the term "centre of gravity" would be quite inappropriate in cases similar to that under consideration.

maintained in its elliptical and nearly circular orbit by the gravitational attraction of the Earth.[1] The Moon must therefore attract the Earth with a force equal to this; and the Earth, being in no way anchored in space, must move under the influence of the Moon's attraction. The fact of its motion is beyond doubt, both from theoretical considerations and from practical observation; and the nature of it is expressed by the statement that the Earth and Moon continually describe similar elliptical and nearly circular orbits about their centre of mass, this point being in a common focus and nearly in the centre of each orbit. The general statement that the Moon describes an elliptical orbit round the Earth is, therefore, though not inexact, incomplete. It would be equally true, and not inexact, to regard the Earth as describing an elliptical orbit round the Moon. Since, however, the mass of the Earth is eighty times that of the Moon, the centre of mass of the pair is eighty times nearer to the centre of the Earth than to the centre of the Moon, lying in consequence well within the Earth itself; so that the actual orbit described by the Moon is far larger than that described by the Earth. It is the common centre of mass of the Earth and Moon that describes an elliptical orbit yearly about the Sun.

With the assistance of the diagram given in fig. 2 there will be no difficulty in forming a definite picture of the system of Algol and its companion, and of

---

[1] The reader may be reminded that the circle is merely a particular form of an ellipse, that in which the greatest and least lines drawn through the centre, or the major and minor axes, are of equal length. The focus of a circle and its centre coincide.

their relative movements. The point o is the centre of mass of the pair, and since it is represented as one-half as far from Algol as from the companion, the companion is regarded as possessing one-half the mass of Algol. The orbits are represented as circles, though, in accordance with the law of gravitation, they might be any variety of similar ellipses. Whatever the relative masses of the pair, and

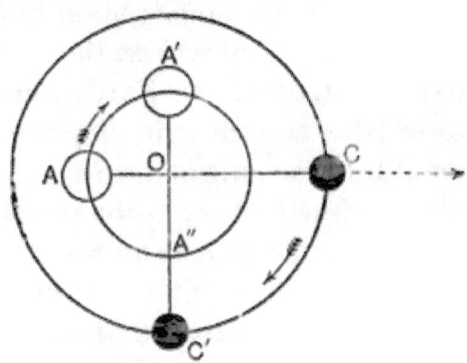

Fig. 2.—The System of Algol.

whatever the degree of ellipticity of the orbits, the general principle, however, remains unaffected. The Solar System is imagined as lying far away upon the right; and, from the fact that no parallax has been detected in Algol, it follows that upon the scale according to which the orbit of Algol is represented, the distance of the Solar System must be, at the least estimate, 5 miles. When at A, Algol will be eclipsed by the interposition between it and the Earth of its dark companion at c. From these positions the star and its companion will proceed in their orbital revolutions, moving in the directions indicated by the arrows, their relative speeds being determined by the condition that the pair must at every instant lie upon opposite sides of the centre of mass, the position of which remains unaffected by their motion. It will be clear, therefore, that if the hypothesis of the dark star's existence is sound, to an observer upon the Earth provided with sufficiently delicate means of observation, Algol should appear to swing to and fro about the point O, attaining its greatest displacement upon either side of it when at A' and A". If, however, the orbit of Algol were even to equal that of the Earth round the sun in magnitude, the consequent displacement in its position would be so slight as to escape detection by the most refined observational means existing,[1] and it has, in fact, never been detected.

During the orbital revolution of Algol there is, however, a relative displacement of another kind between it and the Earth. In executing one half of its orbit the star must continually approach the Earth, while during the other half it must recede from it. Assuming the orbits to be circles, the star should approach the Earth in moving from the position A in which it is eclipsed, the approach becoming direct, and therefore most rapid at A', a quarter period later; while at A" there should exist an equally rapid and direct motion of recession. It is in the detection of these alternate movements of approach and recession that Vogel has achieved one of the most remarkable triumphs of observational astronomy.

The immediate principle, to the successful application of which Vogel's demonstration of the motion of Algol is due, will be more fully explained in a later chapter. It follows as a necessary con-

---

[1] This conclusion is directly involved in the statement that the parallax of Algol is inappreciable by the most refined observational means existing.

sequence of the wave theory of light that a source of light approaching the observer should crowd together and thus shorten its light-waves in front of it, and in consequence alter the nature of the light, raising its colour in the spectral series, that is, causing it to approach the violet in hue, and increasing its refrangibility. A movement of recession should correspondingly draw out, and thus lengthen, the light-waves travelling behind the source and towards the observer, lowering the colour towards the red of the spectrum, and decreasing the refrangibility. In 1842 Doppler had suggested that the apparent colours of certain stars might thus be affected by their movement, a rapidly approaching star acquiring a bluish, and a rapidly receding one a ruddy tinge. The suggestion, however, failed, for various reasons; among them, the fatal one, that, owing to the high speed of light, such transcendent velocities as would be necessary to produce such a change in the colour of a star that should be appreciable to the eye would transform the whole aspect of the heavens in a few years. The true direction in which to search for a record in the light of a star of indications of its approach or recession was indicated by Fizeau in 1848, and lies in the careful measurement of the positions of the dark lines with which the spectra of the Sun and of the greater number of the stars are ruled throughout, the suggestiveness of which was at that time beginning to be recognized. These dark lines simply indicate colours absent in sunlight and starlight, and the absent colours are affected by the motion of the source precisely as are those actually present. Consequently the approach of a star should raise the colours absent in its light towards the violet, and the dark lines in its spectrum should therefore be displaced toward the violet end of the spectrum; the reverse occurring in the case of a receding star. In 1868 Sir William Huggins succeeded in detecting slight displacements in the spectral lines of certain stars, and, assigning the displacements to this cause, estimated from them the velocities of the stars in the direction of the line of sight. It was this method that Pickering suggested should be brought to bear upon the problem of Algol and its hypothetical companion.

In photographs of the spectrum of Algol, taken at intervals during the years 1888 to 1891, the movement of the star, and precisely such movement as was demanded by the eclipse theory, was established beyond doubt. When under eclipse, as well as later by an interval equal to one-half of that between successive eclipses, at which time the star should be between its companion and the Earth, the spectrum of its light should be normal, since at these instants its motion should be directly across the line of sight, and should neither be towards nor from the observer. During the half-period preceding eclipse the motion of the star should be from the observer, and the spectral lines should be therefore displaced towards the red as the result of the drawing out of the light-waves, while after eclipse the motion of recession should be replaced by one of approach, and the spectral lines should be shifted towards the violet. Every one of these predictions was confirmed in Vogel's photographs. The maximum displacement of the lines, which occurred, as they should have, at quarter-periods before and after eclipse, indicated velocities of recession of 24·4, and approach of 28·6 miles per second respectively, the difference between the two values being naturally explained upon the assumption that the speed of Algol in its orbit is 26·5 miles per second, the mean of the two, and that the system of Algol and the companion is approaching the Solar System with a speed of 2·1 miles per second. Knowing the orbital speed of Algol, as well as its period of revolution,—the interval between successive eclipses,—it is a simple matter to calculate the circumference, and from it the radius, of its orbit. The final result is almost exactly a million miles, from which it follows that the oscillation of Algol across the line of sight is far too small to be capable of detection.

The dark companion of Algol suggests the picture of the death-stage of a sun. In nine other stars, variation in light, in nature similar to that exhibited by Algol, points strongly to a similar cause. In another star, Spica, the existence of an invisible companion is indicated by the displacement of spectral lines, though no eclipse results, probably from the plane of the orbits making a sufficiently large angle with the line of sight for the dark star to clear the bright one at each revolution.

In the few cases in which the existence of dark stars has been revealed, their detection has been due to the fact of their close association with bright stars. Only by an inconceivably remote chance would it be possible to become aware of the existence of an isolated dark star by either of the methods that have been so successfully applied to the companion of Algol. Such knowledge of stellar distances as we possess renders it probable that the suns of space are separated from their nearest neighbours by depths so vast that were there dark stars scattered at random among them exceeding the bright ones by many times in number, the probability of one of them approaching so near to a visible star as to sensibly affect it by gravitation would be excessively remote,[1] while, since it is not possible to continually examine more than an insignificant minority of the visible stars, either as regards position upon the face of the sky, by change of which motion across the line of sight would be apparent, or spectroscopically, by which motion in the line of sight might be revealed, it is probable that millions of near approaches between isolated dark stars and brilliant ones would occur before the effect of one would be detected.

The probability of becoming aware of the existence of a dark star by its drifting by chance across the line of sight directed from the Earth toward a more distant brilliant one appears equally remote. It is impossible to contemplate even the most crowded regions of the heavens through a telescope of fine quality, and of large light-grasping power,

---

[1] With the exception possibly of the more crowded regions of the Milky Way.

without recognizing that among the myriads of bright points scattered over the field of view there is ample room for the existence of dark stars far exceeding them in number. The brighter only among the stars appear in the telescope as discs of sensible dimensions; but the reader is probably aware that such "spurious" discs result from imperfections in the eye, and from the inherent principles of telescopic construction. Were the telescope and the eye alike perfect, such is the stupendous remoteness of the stars, that, although suns, the nearest of them, even if far exceeding our Sun in magnitude, would appear under the highest magnifying power that has so far been applied to them, as mere specks of light devoid of sensible dimension.

Although the apparent dimensions of stars are far beyond the possibility of detection with the most perfect optical means, it is, however, possible to make a rough estimate of the extent of sky covered by individual stars or by the whole collection. The possibility of effecting this is based upon the fact that the apparent brightness of any surface is independent of its remoteness. If, for instance, the Sun were removed to three times its present distance, the light received from it would, according to the law of inverse squares, be reduced to one-ninth of its present value; but since its apparent size—or the area in the sky covered by it—would similarly be reduced to one-ninth, the apparent brightness of its surface would remain unchanged. If, therefore, there were a number of sun-like bodies in space, each of the same intrinsic brilliancy as the Sun, but differing both in size and distance from the Earth, it would follow that, since the apparent surface brightness of all would be the same, the amount of light received from any one of them would be in direct proportion to its apparent size, or to the sky-surface covered by it.

The total amount of light received from all the stars above $9\frac{1}{2}$ magnitude (visibility to the naked eye terminates at the 6th magnitude) has been estimated by Mr. Plummer. The result, slightly modified in accordance with more recent measurements of the brightness of Sirius, far the most important star of the whole, is given by Miss Clerke as one-eightieth of that of the Full Moon. The ratio of the light of the Sun to that of the Full Moon has been estimated by Zollner as 619,000 to 1, so that the Sun exceeds the total of the stars above $9\frac{1}{2}$ magnitude in their illumination of the Earth by 80 times 619,000, or nearly 50 million times. If, therefore, the assumption be made that the surfaces of the stars are of the same intrinsic brilliancy as the Sun, it follows that the stars cover a portion of the sky equal to one 50-millionth of that covered by the Sun; and since the Sun covers but one 210,000th of the total of the sky, it follows that the stars would cover rather less than one 10-billionth.

It was necessary to limit the above estimate to the 324,000 stars above the $9\frac{1}{2}$ magnitude as there is no means of determining the light received from the fainter ones. These 324,000 stars, however, far more than include those among which there could be any hope of detecting an eclipse by an isolated dark star.

Should dark stars far exceeding the bright ones in number exist in celestial space, an eclipse of one of the latter would therefore be a phenomenon of rare occurrence; while, should an eclipse occur, so few are the stars the brightness of which is subjected to continual scrutiny that the probability of its passing unnoticed is overwhelming.

The possibility of an unseen system of stars permeating the seen is beyond doubt. The system of the seen is indeed sufficient to satisfy the highest ambition of imagination, but he would be bold who should assert that it may not well form but an insignificant fraction of a still more surpassingly transcendent whole.

The question as to the nature of the changes now taking place in the Sun is one of very great interest, and its study involves physical considerations of high importance and not free from grave difficulty.

The delicate and mottled tracery visible under the most favourable conditions for telescopic observation over the entire surface of the Sun, is strongly suggestive of the view that its bright surface—generally known as the photosphere—consists of an accumulation of incandescent clouds. Such a cloudy structure of the photosphere is in harmony with the general results of solar observation, more especially, perhaps, with the nature and the rapidity of the changes frequently characteristic of sun spots, which, according to this view, are either depressions or actual gaps in the photosphere. The spectroscope indicates as existing above the level of the photosphere a solar atmosphere, as constituents of which the vapours of hydrogen, calcium, iron, and other metals are conspicuous, and in which are traceable with greater difficulty those of a few of the non-metallic elements. It is not difficult to imagine the process of formation of the clouds of the photosphere from the precipitation as fog of the more readily condensable of these vapours by their cooling consequent upon their being carried into the upper regions of the atmosphere.

The question as to the physical conditions existing in the interior of the Sun is attended with graver difficulty, and is of the first importance in the problem under consideration. Herschel imagined as existing beneath the clouds of the photosphere, a solid globe; and even advanced the view, so preposterous to modern students of physical science, that it might, from the protection of a second and intervening cloud shell cool and impervious to heat radiation, be protected from the intense glare of the photosphere above to such an extent as to be a cool and habitable world. When the necessity for the interior heat of the Sun being at least as high as that of its exterior became recognized, the solid globe was generally replaced by an ocean of molten matter.

It is, however, scarcely possible to regard as existing in the interior of the Sun, matter in either the solid or in the liquid condition. The temperature above the photosphere is such that iron, carbon, and other among the most refractory of elements known to terrestrial chemistry are found in it in the gaseous state; and the temperature of these external regions must be far lower than that of the interior. It was for a time regarded as barely possible that the enormous pressure that must exist at great depths in the interior of the Sun might be effective in maintaining matter in the solid or liquid condition in spite of the high temperature, since it is a familiar fact in laboratory experience, that liquefaction of a gas is in every case assisted by pressure, and may in many instances apparently be effected by it alone. Since, however, it became apparent from the classical researches of Dr. Andrews in 1869, that there exists for every element a critical temperature, above which it is impossible for it under any conditions of pressure to assume the liquid state, it has generally been regarded that a liquid interior to the Sun is next to an impossibility. The Sun is, in all probability, essentially an enormous bubble, enveloped in incandescent cloud, from which, by the mechanism of radiation, its energy is transmitted into external space.

From the fact that the degradation of a star from its earliest nebular to its dark state is the direct consequence of the radiation of its heat into space; and as, in ordinary experience, loss of heat is accompanied by fall in temperature, it has frequently been assumed that the life of a star must be the record of continual fall in temperature; and that in the nebulæ would be found the highest temperatures represented in celestial bodies. To the "fiery mist" from which Laplace, in 1796, had imagined the development of the system of the Sun and planets, a temperature was assigned far higher than that of the Sun at present; and the same view was extended to the nebulæ, when the demonstration of their gaseous nature had indicated them as fulfilling in the Cosmos the functions of embryonic stars.

Recent considerations based upon the experimentally ascertained properties of gases and upon the principle of conservation of energy have, however, shown that this simple view cannot be maintained. Attention has already been directed to the fact that each step in the radiation of heat from the Sun brings about a shrinkage of its bulk, or, more exactly, enables the gravitation of its parts to draw them closer together, and that by this act of compression other heat is developed. In a very remarkable paper, published in 1870, Mr. Homer Lane has shown that if the Sun were entirely gaseous, and if the gases composing it were under such physical conditions that the laws of "perfect gases" should be applicable to them, the heat developed by shrinkage must not merely equal but must so far exceed that radiated to effect it, that the temperature of the whole must actually rise in consequence, and must continue to do so for so long as a perfectly gaseous condition is maintained.

A "perfect gas" is defined as one in which, for so long as its temperature is unchanged, any increase in pressure brings about a proportionate decrease in volume. This condition, known as "Boyle's Law", is very closely fulfilled by hydrogen, oxygen, and nitrogen, as well as by gases in general when under conditions far removed from those under which they assume the liquid condition, so long as their density is not rendered excessive by intense pressure. Under extreme pressure, however, decrease in volume becomes increasingly less than that demanded in Boyle's Law, and it is probable that for every gas at a given temperature there is a limiting volume beyond which it cannot be compressed by any pressure however great.

The statement of Lane's theorem—that it is possible under certain conditions for a body to rise in temperature as the result of its loss of heat—appears at first so contrary to common experience that there is generally great difficulty in thoroughly accepting it. That emission of heat is not inseparably associated with fall of temperature, will, however, be clear from the consideration of such instances as are supplied by the condensation of a vapour and the solidification of a liquid. The passage of steam into water at its boiling-point is unaccompanied by any fall in temperature, though the amount of heat given out is more than five times that necessary to raise the temperature of the water from its freezing-point to its boiling-point. Similarly, the freezing of water is unaccompanied by any fall in temperature, though here again a large amount of heat is emitted by the solidifying water—four-fifths of that required to raise the water from its freezing- to its boiling-point. In a mass of gas subject to no external force, development of heat results from its compression under forces due to the gravitation of its parts; it is loss of heat, not fall in temperature, that enables the gravitational forces to become effective in producing compression.

In the hope of assisting the reader towards forming a clear picture of one of the most remarkable of natural processes, a confessedly incomplete demonstration of Lane's theorem is given in the following paragraph, which may be omitted if the mechanical and geometrical principles involved should not appear sufficiently familiar. An essentially similar demonstration, by which indeed the one given here was suggested, is given in Newcomb's *Astronomy*.

Let a globe of a "perfect" gas be imagined, temperature being uniform throughout it, and let the whole be at rest, free from internal currents, and subject only to its own gravitation. All portions of the globe are attracted toward the centre, and a pressure is produced thereby that continually increases toward the centre. According to Boyle's Law the density must similarly increase toward the centre. Let the whole globe be imagined as consisting of a number of concentric spherical shells, each enveloping those within it in the manner suggested by the coats of an onion, and let attention be directed to one of these shells. The total pressure of the gas comprising the shell is due to the weight—that is, the gravitation toward the centre—of the portion of the gas outside of it, and this pressure is distributed over the outer surface of the shell. Now imagine the globe to lose heat by radiation, and to shrink in consequence until its radius has become reduced by one-half. If the process occurs so gradually that the temperature changes uniformly throughout the whole, all portions will shrink equally, the radius of the shell will be reduced to one-half, and therefore, by elementary geometry, its surface will be one-fourth and its volume one-eighth of their former values. The distance of every part of the globe from the centre will be halved and the attraction of each portion to the centre will, since gravitation is inversely proportional to the square of the distance, therefore be increased fourfold. The whole weight of the portion of the globe that lies beyond the shell will therefore be increased fourfold, but, as this weight is now distributed over one-fourth of the former surface, the intensity of the pressure will be increased to sixteen times its former value. Such an increase in the intensity of the pressure would, if the temperature of the shell had remained unchanged, compress the gas in it to one-sixteenth of its former value. It has, however, been shown that the gas occupies one-eighth of its former volume,—double the volume that it should occupy had the temperature remained unchanged. Such an excess in volume can only be due to increased temperature, and its temperature must consequently have risen.

It is scarcely necessary to add that a shrinkage of the radius to one-half is assumed only for the sake of simplicity; the same result—a necessary rise in temperature—would follow from the assumption of any given contraction.

If, then, the Sun behaves as a perfect gas, its temperature must be increasing as the indirect result of the torrent of heat continually radiated into external space. There is good reason to regard it as probable that the Sun is in the main gaseous, but it would be rash to assume that under the extreme conditions of pressure and temperature existing in its interior, the laws of perfect gases are fulfilled by it even approximately. The properties of gases become markedly modified even at such moderately high temperatures and pressures as it is possible to produce in the laboratory, and in such a manner as to suggest that could the matter of the interior of the Sun be subjected to examination, although it would prove to be neither solid nor liquid, it would be difficult to trace in it the gaseous characteristics with which we are familiar. It is therefore impossible to decide the interesting point whether the Sun is at present rising or falling in temperature, though there can be little doubt that in the remote past, when far more tenuous, its temperature must have been lower than it is at the present time.

Whatever the present trend of the temperature of the Sun, it is, to say the least, unnecessary to assume a high temperature for the nebula from which it has been derived. Imagining a nebula from which a single star is to be evolved as a comparatively cool diffuse extension of gas, of so low a temperature and of so great a tenuity that it should obey the laws of perfect gases, not necessarily sufficiently hot for the whole of its constituents to exist in it in the gaseous condition, but possibly embracing them in its volume as discrete solid or liquid particles, it becomes possible to take a rough forecast of its future career. Under the influence of the gravitational attraction upon each other of all its parts, it would tend to acquire a spherical form. Heat would pass into space by radiation; gravitation would in consequence be enabled to draw its parts closer together; the temperature would rise, and portions of its solid or liquid ingredients would become gas.[1] The process would continue, and after a time the nebula would become in the main gaseous. At some period excessive local cooling in the outermost parts would cause condensation there and a photospheric cloud shell would be formed. The nebula has now become a sun. For a time its temperature continues to rise and its radiation becomes more and more effective. At length, however, possibly owing to excessive condensation in the photosphere, possibly to temperature and density increasing in the interior to such an extent that the gaseous laws are widely transgressed, the rise in temperature ceases and is soon replaced by a fall. The Sun has passed the zenith of its career and is now descending towards extinction; a few more ages and its radiant activity has ceased to be.

The question whether any evidence is supplied by stars as to the course they have run from their

---

[1] No doubt heat would also be developed from collisions between the non-gaseous constituents of the nebula, since these would be in motion under the influence of gravitation. It might even be that the first evidence of luminosity in the whole might be due to the generation of heat by these collisions, the gas itself being generally below the temperature of incandescence, as is suggested in Sir Norman Lockyer's *Meteoritic Hypothesis*.

nebulous condition—whether among those visible it is possible to recognize individuals in the early period of their career, others in the meridian of their glory, and others again upon the descending path towards extinction—is among the most fascinating of the speculations of modern astronomy. It is generally regarded that such evidence is indicated in the spectroscopic analysis of their light, but it must be confessed that this branch of scientific inquiry can hardly as yet be regarded as having passed beyond the speculative stage. From it we may hope, perhaps, in the future, to be able to decide whether our Sun is increasing in splendour, or whether he has passed the period of his greatest glory. Here it may be permissible to add that, in the judgment of the writer, the evidence, though not free from serious difficulty in its interpretation, appears to indicate the former as the more probably true hypothesis, and that in the remote future it is not inconceivable that radiations of the Sun should rival even those of Sirius at the present time. Should this be so, the maximum of vitality of a star must be thrown far forward in its life history, and the duration of its decay must be correspondingly brief.

The physical universe is inexpressibly glorious; and it is scarcely possible that the contemplation of the decay of its activity should be unaccompanied by a touch of sadness. One is, therefore, led to inquire, whether among the processes of nature no means are indicated by which its lost energy may be restored to a dead star. So far as the working of nature is revealed in the laws of physical science, the only way in which a star can re-assume its nebulous condition is by a collision between it and another, by which encounter the whole or part of the total energy of motion of the pair would be transformed into heat. The establishment of the equivalence between heat and motion, one of the noblest achievements of modern science, is now a familiar fact to everyone. By the destruction of motion heat is generated; the amount of heat is directly related to the masses and velocities of the moving matter and can be readily calculated from them; while, in its turn, the heat itself may under suitable conditions disappear, and in so doing regenerate motion identical in amount with the quantity that passed out of existence in the act of heat creation.[1]

That many stars are moving relatively to each other is a matter of ready demonstration by observations of their positions upon the sky, with the instruments of refinement now in use, at intervals of a few years. Their movement may be, and commonly is, so apparently insignificant, that centuries must elapse before their displacement would be apparent to the unaided eye; but, upon allowing for their excessive remoteness, speeds are revealed, many

[1] To the term motion, a somewhat vague one as used generally, science applies a definite meaning, the product of mass into velocity. The function of a moving body that is in direct proportion to the heat developed in the alteration of its speed is, however, not this quantity, but the product of its mass into the *square* of its velocity, a quantity to which the term *vis viva* was formerly applied. One-half of the *vis viva*, which is of course also proportional to the heat equivalent, is known in modern mechanics as *kinetic energy* and is of great importance.

comparable with, and some far greater than, those of the planets in their orbits. Sirius drifts over the face of the sky with such speed that in 1400 years its position will be removed from its present one by a distance that would just be covered by the diameter of the Full Moon. From the known distance of the star it is a simple calculation that to do this it must travel athwart the direction of vision with a speed of over ten miles per second, more than one-half of that of the Earth in its orbit; and this takes no account of any velocity the star may possess in the direction of the line of vision, a displacement in which direction would obviously not affect its position upon the face of the heavens. The parallax of Arcturus is inappreciable, from which it appears improbable that its distance can be less than 4,000,000 times that of the Sun; thus remote, the drift of the star, by which it would be carried across the diameter of the Full Moon in 700 years, must represent a velocity of at least 130 miles per second across the line of sight; the actual speed in this direction being greater than this, in direct proportion as the actual distance of the star exceeds the minimum limit that is here assigned to it. From similar considerations it appears, that in the case of a remarkable star in the Great Bear invisible to the naked eye, and known as Groombridge 1830, from the number assigned to it in Groombridge's catalogue, the speed by which the star would be carried in 257 years over such a portion of the heavens as would be covered by the Moon, the most rapid displacement known, must at the distance of the star of 2,300,000 times that of the Sun, indicate a continual rush across the line of sight of 227 miles per second.

The velocity of the Sun relatively to the stars, or, more definitely, the velocity of the Sun relatively to the mean positions of the stars, a quantity commonly alluded to as "the velocity of the Sun in space", an expression almost humorously devoid of meaning, can be estimated from an accumulation of such results as have been here illustrated. The problem was first attacked by Sir William Herschel, and has ever since been a favourite matter of research of astronomers, who have been enabled to introduce increasing refinements as more and more data have become available. All the methods that have been applied consist essentially of the determination of the average velocities of the stars, that is, the determination of the velocity of the average position of the stars relatively to the Sun, that of the Sun relatively to the mean position of the stars being equal and opposite to this. The outcome of such investigations seems to indicate that the Sun is travelling in a line directed very nearly towards the brilliant star Vega, and that its velocity in this direction is probably between 12 and 18 miles per second. There is no doubt that the result as regards direction is far more definite and accurate than that as regards speed.

In the host of the "fixed stars" is found abundance of motion, and that upon the most stupendous scale. A century ago it was fondly hoped that the movements of the stars might turn out to be of the orderly and permanent character revealed in the Solar System, and search was made for a colossal Sun, that should by its gravitational attraction control the whole. Sirius was suggested by Kant, other stars took its place in succession, and in 1846 Mädler, abolishing the conception of a central Sun, imagined that every member of the stellar host might describe an orbit about a centre, placed by him in the Pleiades, the controlling power being, not the overpowering attraction of one, but the combined influence of all. As the motions of the stars became more closely followed, it became clear that the hope of revealed order was not destined to be realized. System remains unrevealed in their movements, and the stars appear to rush in random directions through space.

The problem before us then is, whether in their undirected career stars may not from time to time come into collision. Were the Earth in its orbital speed to meet in direct impact another planet, equal to it in mass and travelling with an equal speed in the opposite direction, and were the planets to escape being shattered into fragments by the impact, heat would be developed from the destruction of their motion sufficient in quantity to convert both into a cloud of gas,[1] and it is conceivable that

---

[1] The collision between two solid planets might result in the shattering of considerable portions of them into fragments, and in the fragments being projected into space with high velocities. The motion retained by these fragments would, of course, escape being converted into heat. In the case of stars that had not cooled so far as to reach the solid condition, such shattering would be less probable. See a paper by Lord Kelvin, *Popular Lectures and Addresses*, vol. i. p. 366.

a like result might arise from collision between stars. From the insignificant dimensions of the visible stars in comparison with the celestial spaces in which they have their being, the chance against a collision, even in geological ages, is perhaps excessively remote; but in indefinitely prolonged time collision appears certain. It must be remembered, in addition, that the stars that are seen may well be but a small fraction of the whole system; and with each addition of dark suns the probability of collision becomes more than proportionately greater.

Regarding, however, the rejuvenescence of a star by collision as possible, the last catastrophe is but projected forward by a finite time. At each collision the coalescence of a pair of cosmic masses will reduce the existing number by one; while energy of heat is gained at the expense of energy of motion. As æon succeeds æon, and as new nebulæ follow those from the ruins of which they were formed into extinction, the Universe becomes poorer in active energy; and there appears, so far as physical science has interpreted the processes of nature, no escape from the picture of an accumulation of inert matter as the last memorial of a glorious Universe of Suns.

Appendix to Chapter I.

The Measurement of Stellar Distances.

Bessel's discovery of stellar parallax, a discovery that directly demonstrated the fact of the Earth's annual motion round the Sun, has been generally regarded as the first direct proof of the truth of the Copernican System of Astronomy; though without doubt a very strong case for priority in this respect might be made out for the detection by Bradley, rather more than a century previously, of the aberration of light. In any case, however, Bessel's achievement removed the last and a very serious objection to the Copernican Hypothesis however firmly established, and has rendered it in every respect unassailable. The discovery itself must take high rank among the greatest triumphs of observation. The mere detection of so minute an angle as even the relatively large parallax of 61 Cygni still necessitates instrumental means of extreme refinement, as well as very great observational skill. Two lines inclined at an angle of a second of arc would approach by no more than 1 inch in a distance of $3\frac{1}{4}$ miles, and the inclination, not only detected by Bessel, but measured with considerable accuracy, was but a fraction of this. If the smallness of the angles concerned were the only difficulty in observations of stellar parallax, its detection would be no mean feat; but the observations are affected by numerous sources of error, the elimination of which involves the utmost perseverance. The necessary observations must, of course, be made at widely separated times of the year, and difference of temperature not unfrequently causes change in the form and the position of the observing telescope, that would, if not taken into account, completely conceal the insignificant angle of parallax by simply overwhelming it. From a principle similar to that by which the rain-drops of a falling shower appear to slant towards a moving passenger to an extent dependent upon the rapidity of his motion, light rays arriving from a star appear to an observer upon the moving Earth—as he is carried by it in its orbital rush across their streams—to slant from the direction of the Earth's motion, and the star appears, in consequence, to be displaced toward that point in the heavens to which the Earth's motion is at the time directed. The phenomenon is known as "aberration of light"; it was indeed discovered by Bradley in 1725 during an unsuccessful attempt to detect the parallax of a star, and the displacement in the apparent position of a star due to it varies from nothing to 20 seconds of arc according to the direction of the Earth's motion with respect to that of the star. In estimating the true direction of a star from its apparent place in the sky it is obviously necessary to take the most careful account of the aberration of light.

The apparent position of a star is also seriously affected by refraction. In accordance with the general fact that a ray of light is deflected, or refracted, in passing from a medium into another differing from it in density, the refraction being towards the perpendicular to the separating surface as the ray passes from a rarer into a denser medium, the rays from a star, after following a straight course in external space, are deflected downwards on entering the atmosphere, and as the air continually increases in density as the surface of the Earth is approached, the deflection continually increases, so that the ray reaches an observer after executing a curve in the atmosphere, and the apparent direction of the star, determined by the direction of the ray on entering the eye, is sensibly different from its true direction. Unlike the interference caused by aberration, which may be corrected from an exact knowledge of the speed of the Earth relatively to that of light, the error due to refraction is incapable of exact determination, since the curvature of the ray is dependent upon the density, temperature, and degree of moisture of the air, not only in the observatory, but throughout the whole of its atmospheric path, which is quite beyond the reach of observation.

With a view of minimizing, and avoiding as far as possible, these and certain other sources of error, Bessel adopted in his observations upon 61 Cygni a method originally proposed by Galileo, and known as that of "relative parallaxes". In it no attempt was made to determine the absolute direction of the star with exactitude, but its direction was determined with a very high degree of accuracy with reference to a neighbouring "reference star",

The Measurement of Stellar Distances.

and the observations were repeated at different times of the year. The important assumption was then made, and its justification will be examined presently, that the reference star was so extremely remote that the direction in which it was seen was not appreciably affected by the Earth's movement—in other words, that it possessed no parallax that could be detected —its apparent proximity to the star under observation being merely the result of its lying by chance nearly in the same direction. The principle underlying the application of the observations will be clear from the diagram of

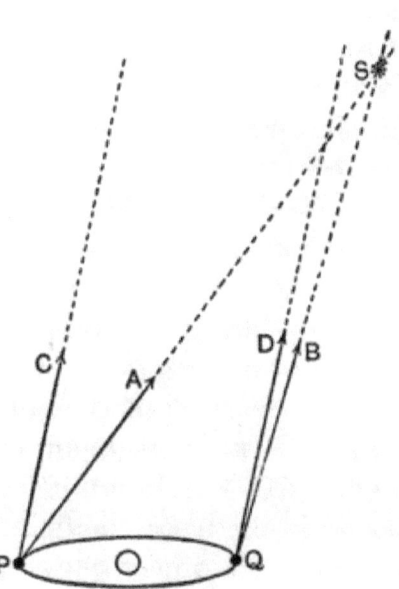

Fig. 3.—Relative Parallax.

fig. 3. Let the two parallel lines PC and QD represent the lines of sight from the earth in its two positions P and Q to the reference star, assumed to be so remote that there is no inclination between them capable of detection. Then, the Earth being at P, let the direction of the star under examination be observed with reference to the reference star, by measuring the angle between them. Setting off this angle in the figure as CPA, the direction of the star is determined by the straight line PA. Six months later, the

Earth being at Q, a similar observation may be made; if the angle separating the stars be observed, and set off as DQB, the direction of the star is now indicated by the straight line QB. The inclination of PA and QB determine the position of the star at their meeting point at S. It need scarcely be remarked that in actual practice the position and distance of the star are determined by trigonometrical calculations based upon the observed angles, and not from graphical construction such as has been here introduced to illustrate the principle involved.

The instrument that has so far been found best adapted to the measurement of the required angles is known as the heliometer, from the fact of its having been originally designed to determine the angular measure of the Sun's diameter; and it was with such an instrument, constructed by the celebrated optician Fraunhofer of Munich, that Bessel's observations were made. The heliometer is a telescope, the object-glass of which is cut into two along a diameter. With the two halves in their normal positions each may be regarded as giving a separate image of an object towards which the instrument is directed, but the pair of images coincide, and under these conditions the heliometer is equivalent to an ordinary telescope. If, however, one of the halves is displaced by sliding it along the line dividing the pair, the image formed by it is equally displaced, and the observer at the common eye-piece sees all objects in the field of view doubled, as if viewed through a crystal of Iceland-spar. In practice one half of the object-glass is displaced until the image of the star under observation for parallax formed by it coincides with the image of the reference star formed by the other. The amount of displacement between the two halves of the object-glass then indicates the angle between the directions of the stars.

The advantage, as well as the one grave disadvantage, inherent in the method of relative parallaxes will be clear upon consideration. The angles CPA and DQB are not appreciably affected by aberration, since, the pair of stars under observation lying in nearly the same direction in space, their rays will traverse closely coincident paths, and the apparent change in their direction caused by the Earth's rushing across the streams of light rays will be therefore nearly the same for each. For a similar reason, error due to refraction is practically eliminated, since the rays from the two stars traverse nearly the same column of atmosphere, and are therefore refracted by it almost equally.

It is important to notice that it is not necessary to know the absolute directions of the stars with great accuracy. So long as PC and QD, the lines of sight to the reference star, may be regarded as parallel, a displacement of the pair of them to and fro even through several degrees makes quite an insignificant change in the estimated distance of the star at S. It would be instructive to verify this by repeating the construction for directions of these lines slightly different to those in the figure, being careful to represent them as parallel in every case, and to retain the exact values of the angles CPA and DQB, which are those determined by the heliometer.

The one serious and obvious objection to the method of relative parallaxes lies in the necessity for assuming that the reference star is so remote that its parallax is inappreciable. It will be readily seen, by modifying the construction of the figure, that if the lines of sight to the reference star are inclined to each other, the distance of the star s will be underestimated. The justification for the method may perhaps be stated in some such way as the following. The enormous majority of the stars show no parallax relatively to each other that can be detected. This must arise, either from all such stars being so immensely remote that their parallaxes are represented by angles so small as to be beyond the power of appreciation even by the heliometer; or, if the stars are so near the Earth that these angles are appreciable, from their all being equally remote, so that all experience equal apparent displacements when regarded from different points of the Earth's orbit. There can be no hesitation in regarding the first alternative as overwhelmingly probable. There is, therefore, a great probability that any given star selected is sufficiently remote for its parallax to be ignored in its use as a reference star; and if concordant results are obtained from the employment of two reference stars, which should always be the case for the work to inspire confidence, the probability of the soundness of the assumption becomes overwhelming.

Before the application by Bessel of the method of

relative parallaxes, the attempt had generally been made to determine the absolute direction of the star under observation for parallax by recording its position upon the face of the heavens without the assistance of reference stars. The position of the star was determined by the observation of its "right ascension" and "declination", celestial quantities quite analogous to longitude and latitude upon the surface of the Earth. Such a method is known as that of "absolute parallaxes", and it is preferable to the relative method in that it does not involve the aid of reference stars. Owing, however, to observations by the absolute method being affected to the fullest extent by refraction, aberration, and other troubles, as well as from its involving the measurement of large angles, which are far more difficult of determination within the same limits of absolute error than small ones, the possibility of its successful application, even to the mere detection of parallax, in any given case, is so extremely remote that it has now been universally rejected. It is true, Henderson's discovery of the large parallax of α Centauri was effected by the absolute method, but success could be scarcely anticipated with parallaxes of much smaller value.

Bessel's success was therefore largely due to the adoption by him of the method of relative parallaxes, as well as to the fact of his being in command of a fine heliometer. It was also largely due to the judicious selection of a star for examination. Previous to Bessel's measurements there was reason for regarding is as probable that 61 Cygni was one of the nearest of the stars, and that it therefore possessed a relatively large parallax. In appearance one of the most insignificant and unattractive of the stars, attention had been for some time directed to 61 Cygni by reason of its rapid drift across the face of the sky. So great is this "proper motion" in its case that in 350 years the star would traverse a line in the heavens equal to that covered by the diameter of the Full Moon, a rapidity of movement exceeded, so far as is known, only by two other stars. Since, other conditions remaining the same, the nearer a moving star the more rapidly would it appear to drift across the sky, it is evidently probable that the more rapidly drifting stars are as a class nearer than the others; hence in the search after the parallaxes of stars special attention has been directed to them, and 61 Cygni specially attracted the attention of Bessel. Another criterion of probable nearness is supplied by brightness, it being *a priori* probable that the brighter stars are nearer the Earth than fainter, and for this reason special attention has also been devoted to them. It is interesting to notice that in the result rapidity of motion has proved a far more favourable omen of success in the search after parallax than great brilliancy.

Chapter II.

The Milky Way and the Distribution of Stars.

Among the many and profound problems suggested to the mind by the contemplation of the heavens upon a clear, moonless night, there is no one more mysterious, and few have proved more baffling, than that presented by the dimly-luminous arch of the Milky Way. Variously regarded in classical mythology as the milk that flowed from the sacred breast of Juno; as the last vestige of the ruin that overwhelmed Phaeton in his bold but fatal attempt to direct the fiery steeds of the Sun's chariot; and as the road along which the gods repaired to High Olympus; the fair shimmer of the Milky Way has through succeeding ages been associated with poetic fancy and romantic imagination. In modern German the popular term "Jacobstrasse" recalls not unfitly the sublime vision of the patriarch of Israel; while to the Indian of North America the Milky Way is still the path of departed souls, and the brighter stars that stud its stream are the camp-fires that mark the halting-places of his fathers upon their weary march.

From a very early date there have been recorded speculations of a more or less scientific nature regarding the Milky Way. Pythagoras is recorded to have formed the shrewd conjecture that its faint shimmer was due to the accumulated light of multitudes of faint stars. Anaxagoras maintained the view that the appearance might be caused by the projection into space of the shadow of the Earth; while Aristotle regarded it as a mist formed by the exhalation of terrestrial vapours. That it was a ring of nebulous matter in external space encircling the Earth appeared the probably correct solution to both Tycho Brahe and John Kepler toward the close of the sixteenth century; but a few years later its true character was revealed in Galileo's telescope, and the speculation of Pythagoras was confirmed in the discovery that its haze is indeed the combined shimmer of hosts of stars, each one too faint by itself to be distinguished by the unaided eye.

Seen under the most favourable conditions from these latitudes, the Milky Way appears as a semi-circular arch of light spanning the starlit sky. The appearance suggests that what is seen is the visible half of a complete zone encircling the heavens, the other half being at the time of observation in the celestial hemisphere that is hidden by the solid Earth under foot. On constructing a complete map of the Celestial Sphere, and tracing the course of the Milky Way upon it, it is seen that this view is roughly correct; but the entire stream departs from a simple zone-like character in several respects. For about two-thirds of its circuit of the heavens, the Milky Way, though irregular, appears as an unbroken stream. In the constellation of the Swan, however, it bifurcates, the divided branches, after following appreciably parallel tracts for about one hundred degrees, reuniting in the constellation of the Centaur. The division in the Swan is excellently situated for observation from these latitudes, from which, however, the reunion in the Centaur is, owing to its proximity to the South Pole of the heavens, permanently invisible. The luminosity of the more northerly of the branches of the divided stream fades at a short distance from the bifurcation in the Swan, and, indeed, ceases for some distance, while it is very remarkable that this fading is accompanied by an increased brilliance in the southern branch. That the branches, though separated, are not physically independent is indicated by the existence of imperfect bridges of luminous matter between them, while in several points the more southern and brighter branch throws off comparatively brilliant projections toward its companion, which projections, however, terminate before reaching it.

A few degrees from the permanent reunion of its divided streams, and upon entering the constellation of the Southern Cross, the course of the Milky Way expands into a brilliant and well-defined cloud of stars, while, in the centre of the cloud, closely bordering upon the four bright stars of the Cross, is the dark pear-shaped lake popularly known as the "Coal Sack". The "Coal Sack" is one of a great number of similar irregularities in the Milky Way, though in no other is the passage from extreme richness in stars to almost total vacuity so sudden. The appearance of a dark void, unthinkable in extent, in the midst of a cloud resplendent with the light of tens of thousands of suns, is indeed one of the most impressive features presented by the system of the stars, and it is scarcely remarkable that no explanation of it has been advanced that it is not easy to refute upon the most elementary grounds.

A third and no less remarkable feature of the Milky Way is presented in the same region of the heavens. Following the course of the Milky Way past the Coal Sack, its stream contracts, becoming almost at once reduced to little more than a narrow neck of light. Beyond this, however, it as rapidly widens, and a few degrees farther on, in the constellation of Argo, it is broken clear across by a dark chasm. In Sir John Herschel's beautiful drawing of the Milky Way in the Southern Hemisphere the stream upon either side of the gap is shown as extending finger-like projections of faint light towards the opposite side, as if in vain endeavours to bridge across some invisible barrier, while in Gould's more recent drawing, executed at Cordova, multitudes of faint stars are represented as scattered over the break.

There is but little direct evidence as to the distance of the Milky Way. It is conceivable that the distance of a portion of the Milky Way might be revealed in observations for parallax made upon a star, apparently lying in its stream while actually situated far beyond. In such a case, if a number of Milky-Way stars showed the same relative parallax with reference to the selected star, it would follow that the members of the group were at the same distance from the Earth, and if the parallax of the selected star were inappreciable, which would be probable if other selected stars, similarly examined with proper precautions, gave the same result, the distance of the Milky-Way stars would be determined. No such effect as this has, however, been observed, and there can be little doubt that the Milky Way is more remote than at least many of the stars whose parallaxes have been determined. There is, however, no known reason why the method should not be successfully applied in the future.

At a first casual glance the stream of the Milky Way appears to be of a uniformly diffused luminosity throughout. Upon more careful inspection, however, the first suggestion of uniformity disappears, and a vast amount of varied and intricate structure becomes revealed throughout the entire system. Some of this detail is readily apparent to the naked eye upon a clear moonless night; but, to appreciate it in its full beauty, the keenest sight, supplemented by the most favourable conditions, is essential. Such conditions involve, of course, a moonless night; a highly transparent atmosphere; a time when, as is admirably the case in the autumn and winter evenings, the Milky Way extends high into the vault of heaven, so that the greater part of it escapes the effects of light absorption by the atmosphere, always so appreciable when viewing celestial objects low down towards the horizon; and a locality well removed from sources of artificial light, and the consequent glare that they produce in the sky.

As the eye, under such conditions, becomes more and more accustomed to its faint light, the uniform luminosity at first suggested by the Milky Way becomes replaced by details of structure steadily emerging from its haze, until the whole stream spanning the sky assumes an entirely new signification. Consisting in some parts of clouds of faint stars, separated by connected dark or dusky rifts; in others, of wisps of starry matter, sometimes interlacing in inextricable maze over the body of the stream itself, and frequently projected as delicate streamers far into the neighbouring sky; the whole appearance of the Milky Way has suggested, not inaptly, the image of the knotted and gnarled trunk of an old forest tree. The interpretation of the scheme thus dimly unfolded may be beyond our power, but in surveying it, the observer becomes powerfully impressed with the conviction, that rather than being a fortuitous concourse of suns, the Milky Way is a system possessing a complicated and varied structure.

The telescopic appearance of many regions of the Milky Way is of extreme beauty, and structure is revealed of a more minute character; that appreciable to the unaided eye being not unfrequently lost owing to the limited area of the sky that it is possible to embrace in one view. Appearing in some regions as a collection of individual stars scattered apparently at random over the dark background of the sky; in others as clouds of innumerable stars, which as his telescope moved, suggested to Sir John Herschel the image of a drifting scud; while not unfrequently its hosts of stars appear as if involved in extensive nebulosity; the Milky Way, when viewed through a fine instrument of large light-grasping power,

The Milky Way and Star Distribution. 65

presents a picture, as its "clusters and bee-like swarms of stars" drift in silent procession across the field of view, that is in the highest degree impressive. To the true star-gazer, the whole picture possesses an inexpressible and quiet charm, and suggests thoughts to which few would find it easy to give expression.

During the past twenty years the invention of the gelatine dry plate, and the continual improvements effected in its preparation, have placed a new method of research, and one of tremendous power, at the disposal of the astronomer. It is now a matter of common knowledge that, in its power of recording very faint light, the photographic plate far surpasses eye observation, for the simple reason that, to be appreciated by the eye at all, an object must be of a certain brightness; but that the effect of light upon the photographic plate being cumulative, a perceptible image may be produced by light far below the limit of visibility to the eye, if its action upon the plate be allowed to continue for a sufficiently long time. It is consequently in the representations of extremely faint objects, such as the nebulæ and the streams of the Milky Way, that photography has achieved its greatest triumphs in this direction.

In its application to astronomy, the photographic plate is usually placed within a telescope from which all lenses, with the exception of the object-glass, have been removed; and at a distance behind the object-glass equal to its focal length. Under these conditions a sharply-defined image of the celestial object towards which the telescope is directed is formed by the object-glass upon the plate. In some of the most beautiful studies of the Milky Way, however, a portrait lens and a camera closely resembling the usual form have been substituted for the telescope. Larger pictures are obtained by the former method, but the latter embraces a larger field of view and is the more sensitive. It is scarcely necessary to add that, in either method, the entire instrument must be mounted appropriately and driven by clockwork so as to accurately follow the celestial object in its apparent diurnal revolution round the Earth.

Of the applications of the photographic plate to the study of the minute detail of the Milky Way it will only be necessary to refer to the beautiful work of Mr. Barnard at the Lick Observatory. From the glorious situation of the observatory upon Mount Hamilton, under conditions nearly perfect, so far as atmospheric transparency is concerned; with 11,000 feet, and that the most troublesome portion, of the atmosphere below, Mr. Barnard, using a portrait lens of only 6 inches in aperture, and exposing the plates for periods varying from one to twelve hours, has obtained a series of pictures of the Milky Way of the utmost beauty and delicacy. It is scarcely too much to say that the pictures as far surpass, in the amount of detail revealed, the view presented to the eye through the largest telescope, as does the latter the result of naked-eye observation; though it must be confessed that the point-like images and the brilliancy of the star pictures, both of which add so greatly to the beauty of the telescopic view, are lost in the photograph.

It is impossible to examine the exquisite photographs obtained by Mr. Barnard, even superficially, without being impressed with the sense that over enormous regions of the Milky Way, of a space so vast that in comparison with them the whole of the Solar System would shrink to utter insignificance, influences have been at work, concerning the very nature of which it is only possible to form the vaguest conjecture. The most striking features are perhaps the dark lanes or rifts that so frequently appear to intersect clouds of stars. These rifts seldom occur in isolation; they more generally form branching systems, many branches frequently radiating from a common trunk or other apparent vacuity. In some cases the rifts are dark, with sharply-defined edges, and are nearly, or quite, devoid of stars: in others, they are uniformly hazy, as if viewed through an interposed star cloud: while, again, they appear of dusky and less regular forms that have suggested the view of dark clouds of cosmic dust lying between the Earth and a more distant background of brilliant stars. Another very suggestive feature in the minute structure of the Milky Way is the frequent occurrence in it of stars arranged in lines. The lines of stars are frequently simple, but they often assume curved and branching forms, and a very general characteristic is shown in their tendency to arrangement in directions roughly parallel to dark rifts in the same region. A remarkable case is supplied in the constellation of Sagittarius, where Mr. Barnard's photograph shows a group of upwards of thirty small stars arranged in the form of a forked twig, the end of the twig remote from the fork being sharply curved round into a hook. In the immediate neighbourhood of this group, and generally parallel to it, are several well-marked, dark, and dusky rifts. In many cases physical relationship between stars thus arranged is emphasized by the occurrence of streams and wreaths of nebulous matter connecting and involving the stars. Star streams also frequently assume the form of closed oval curves, the included space being commonly darker than the exterior.

The question whether the Milky Way is an isolated structure in space, or whether it is related in any recognizable matter to the system of the stars, is one that has given rise to much study and speculation during the last hundred years. Towards the close of the last century Sir William Herschel noticed that, although the stars were distributed over the face of the sky with great irregularity, there was upon the whole a decided tendency to increased density in their distribution towards the region of the Milky Way. Confining his attention, by reason of the extended nature of the problem, to a zone of moderate width intersecting the Milky Way at right angles, and directing his reflecting telescope of $18^{7}/_{10}$-inch aperture towards every part, in succession, of more than one-half of it, upwards of three thousand observations were made of the number of stars visible at any one time

The Milky Way and Star Distribution. 69

in the field of view. The area of sky embraced in any one view was nearly one-quarter of that covered by the Full Moon, and, taking the average density of star distribution for equal distances from the Milky Way, it appeared that fewest stars appeared at a distance of 90 degrees from it, and that the number continually increased as its stream was approached. In equal steps of 15 degrees each from the poles of the Milky Way to its middle plane the average numbers of stars visible in the field of view were 4, 5, 8, 14, 24, and 53.

In recent years the crowding of stars towards the zone of the Milky Way has been examined in a more detailed manner. Herschel was content to merely count the total number visible at any one time through his telescope, ignoring any distinction as regards brightness between the stars themselves. In 1870 Mr. Proctor showed, by counting the number of lucid stars, *i.e.* those visible to the naked eye, that a similar crowding was recognizable with respect to them; and still later Mr. Gore has shown that a similar crowding can be traced in stars of each individual magnitude, taken separately. Further, a remarkable law of crowding becomes apparent in treating the problem in this manner. With the brightest of the stars the crowding to the Milky Way is very marked, so much so that, as Mr. Proctor has remarked, upon a moonlit night, when the Milky Way is itself invisible, it is still possible to trace its course, at any rate in the Southern Hemisphere, by the track marked by the brilliant stars. With descending steps in brilliancy the crowding of the stars becomes, however, less emphasized; for stars just upon the limit of vision it is scarcely recognizable, while for telescopic stars just below that limit it practically vanishes. For still fainter stars, however, the crowding is again apparent, and continues to become more and more pronounced until the faintest of the telescopic stars are reached.

It is easy as well as interesting to trace this remarkable law of distribution in the positions of the brightest stars. Of the ten most brilliant stars of the northern hemisphere, three—Capella, Altair, and Alpha Cygni—are situated very near to the central line of the Milky Way, though its entire stream does not occupy more than one-seventh of the hemisphere. Four others—Vega, Procyon, Betelgeux, and Aldebaran—are placed upon its immediate border, and all have been thought to be involved in its faint extensions. A zone embracing four-fifths of the sky, with the Milky Way in its middle plane, contains, in addition to these seven, the eighth, Pollux, that is, exactly twice as many stars as it should if the distribution had been uniform. Two only of the ten, Regulus and Arcturus, are far removed from the Milky Way, and it is suggestive that of this pair, Arcturus is notorious by reason of its very high proper motion, or drift, across the face of the sky; one that is only exceeded in magnitude by eighteen other stars. It is quite conceivable that an enormous speed might enable a star to resist a tendency to distribution to which less rapidly moving bodies would conform.

From the remarkable law of star distribution traced by Herschel and later observers, it appears scarcely possible to regard the Milky Way as an independent and isolated formation; while a definite relation between it and other celestial objects is still further emphasized by a study of the distribution of the nebulæ in space. In the course of his researches upon these celestial clouds Sir William Herschel became aware of a curious antipathy displayed by them towards the brighter stars. He continually found groups of nebulæ in spaces of the heavens comparatively destitute of stars, and separated from the richer regions around them by dark spaces. So firmly did he become impressed with the reality of this relationship of avoidance, that, during the process of "sky sweeping", as his great telescope was directed toward different regions of the heavens, he was accustomed to warn his assistant to prepare to record nebulæ, for, by the thinning out of stars, he anticipated that he was approaching nebulous ground. He observed that the crowded stream of the Milky Way was almost destitute of nebulæ, but that toward its poles, where stars were most sparsely distributed, nebulæ appeared in greatest number. The avoidance of the Milky Way displayed by nebulæ was still further emphasized in the observations of Sir John Herschel, who extended his father's researches into the Southern Hemisphere.

The avoidance of the zone of the Milky Way displayed by nebulæ, and their condensation toward its poles, may be very strikingly shown by constructing a map of the heavens, upon which the positions of the nebulæ and that of the stream of the Milky Way are recorded. Such maps have been frequently made, the most recent being by the late Mr. Sydney Waters, who in 1893 marked, upon the method known as that of "equal surface projection", the positions of the 7840 nebulæ and star clusters recorded in Dr. Dreyer's New General Catalogue of 1888. In view of the fact that at a not very remote date the belief was generally entertained that nebulæ were clusters of stars, it is interesting to notice the relationship displayed by star clusters towards the Milky Way. Their condensation toward it is even more pronounced than that of the stars, very few being found beyond the limits of the Milky Way, while for a considerable part of its course the centre of its stream is occupied by a continuous line of them.

The application of the spectroscope to the study of the nebulæ has brought to light a curious modification of their law of distribution. We have seen in the previous chapter that many of the nebulæ, by yielding a broken spectrum, are thereby demonstrated, at any rate so far as their luminous constituents are concerned, to be in a gaseous condition, but that the light from others yields a continuous spectrum, which is at present incapable of exact interpretation. On examining the manner in which these two classes of nebulæ are distributed in the heavens, it appears that the "gaseous" nebulæ share the tendency towards condensation displayed by stars in crowding towards the Milky Way, the greater number of nebulæ actually appearing in it being gaseous. Upon deducting these from the total number, and mapping the positions of the remainder, the avoidance of the Milky Way displayed by them is consequently more pronounced than before.

In following the steps by which structure has been traced throughout the entire stream of the Milky Way, and system revealed in the distribution of the stars and nebulæ with reference to it and to each other, we have so far followed a sure course, and indeed have had occasion to do little more than record the results of direct observation. That stars and nebulæ are not scattered at random throughout space, that there is law in their manner of distribution upon the face of the sky, and that with this law the stream of the Milky Way is in some manner intimately associated, cannot be doubted. So much is known; and it is scarcely possible to refrain from the attempt to gain some closer insight into the nature and meaning of the entire system. Here, however, our steps at once follow a less certain track, and rapidly lead toward a region of speculation in which not a few sound thinkers have become sadly bewildered.

In 1784 Sir William Herschel, adopting in principle a suggestion advanced by Thomas Wright of Durham thirty-five years earlier, attempted to account for the appearance of the Milky Way, and the condensation of stars toward it, as a perspective effect. According to this view, all stars, including those that appear to form the Milky Way, were assumed to be, upon a broad average, uniformly distributed in a stratum or layer, the thickness of which was small in comparison with its dimensions in its own plane; while the Sun was situated not far from the centre of the stratum. It is clear that according to such a law of distribution the line of sight from the Solar System directed at right angles to the stratum would soon emerge into external space, and that but few stars would appear in this direction; but that in all directions parallel to the faces of the stratum stars would appear to be crowded, since the line of vision would be, in all these directions, for a long distance involved among stars. All round the Solar System, in the middle plane of the stratum, stars would, therefore, appear to be crowded, and by such perspective effect the appearance of the Milky Way was imagined to be produced. In passing from a direction along to one at right angles to the stratum, the length of the line of sight included in it would continually decrease, and fewer and fewer stars would therefore be seen toward the poles of the Milky Way. It was further suggested that the bifurcation of the Milky Way was an optical effect, due to the projection from the principal stratum of a secondary one, making a small angle with it, and leaving it nearly in the direction of a straight line passing from the Sun.

If the stars were distributed with perfect uniformity throughout space, and if the most distant and the faintest were visible through a given telescope, it would of course be possible, by counting the numbers visible in equal areas of the heavens, to compare the extensions of the system of the stars in the corresponding directions. For a time Herschel believed that these conditions were fulfilled with sufficient approach to accuracy to justify the application of the method; and, with a view of "fathoming the Universe" in this manner, he carried out a laborious series of "star-gages", his telescope being directed successively to upwards of 3000 selected regions of the heavens, and the number of stars in each field of view counted. The results were published in 1785.

It follows from simple geometry, that if the stars are distributed uniformly, and if the telescope employed is sufficiently powerful to reveal all of them, the extension of the system in different directions is proportional to the cube root of the number of stars

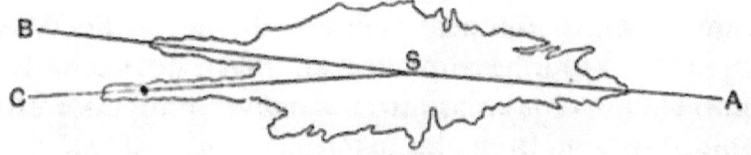

Fig. 4.—Sir William Herschel's earlier view regarding the Form of the Stellar Universe.

appearing in equal areas of the heavens in those directions; and upon this principle Herschel constructed the familiar hypothetical section of the sidereal system which is reproduced in fig. 4. In this the Solar System occupies the position indicated by the point s, the straight line sA is directed rather to the north of Sirius, and the great length of it involved among stars gives rise to an optical crowding to which the appearance of the Milky Way is ascribed; the cleft upon the opposite side accounts upon the same principle for the apparent division of the Milky Way into two branches, the lines SB and SC, by fathoming the starry extensions shown in the section, giving rise to the bifurcation in the neighbourhood of Altair.

To the reader, acquainted with the intricate structure of the Milky Way revealed by more recent observation, it will be at once apparent how completely at variance with nature was Herschel's assumption that its appearance could be due to any such merely optical effect; but it is difficult to imagine how, with the knowledge of star distribution that was the result of his own work, Herschel could have, even for a time, regarded his fundamental assumption of approximately uniform distribution as valid. To account for the Coal Sack and the other less-pronounced vacuities of the Milky Way upon the hypothesis of optical effect, vast cone-like tunnels must be imagined as converging upon the Solar System from the external void; while every appearance of exceptional richness must be regarded as arising from immense columns of stars projecting from the system afar into external space, and similarly radiating from the Sun. As a centre of such converging tunnels of vacuity and radiating projections of starry cones, the Sun assumes a unique place in the Universe entirely without warrant. Upon the optical hypothesis, again, except upon similarly extravagant assumptions, the appearance of crowding of stars towards and within the Milky Way should take place gradually. From the external sky to the centre of its stream, density of star distribution should increase continually and by such perfect gradation that it should be impossible to define the limits of the Milky Way. Such an effect is, however, directly contradicted by observation. Throughout its entire course the boundary of the Milky Way is generally marked with fair definition, while in some regions the transition from its star-crowded depths to the external sky is so abrupt that one half of the field of view of the telescope may be dark while the other half is crowded with its stars. Such features point definitely to the Milky Way being a real and definite structure, separate from, though doubtless in some way closely related to, external celestial bodies. That this conclusion is unavoidable was clearly recognized in his later life by Herschel himself, and was fully acknowledged by him, although, unfortunately, his earlier hypothesis was never formally withdrawn.

Closely bearing upon the problem of the Milky Way and the relation of external objects to it, is the question whether the stars are distributed over a finite region, or whether they are scattered without limit through infinite space. In other words, is the visible universe finite or infinite in extent? Upon the random suppositions that stars are distributed uniformly and without limit through infinite space, and that all are of the same magnitude and of the same intrinsic brilliancy as the Sun, a very remarkable conclusion is reached, which is often quoted, and to which it is worth while to direct attention. With the Earth as centre, let an infinite number of concentric spherical surfaces be imagined in space having radii proportional to the natural numbers 1, 2, 3, 4, &c., and let a certain number of stars be imagined to be distributed over the first surface. The second surface, having twice the radius of the first, will have four times the area, and will therefore contain, upon the hypothesis of uniform distribution, four times as many stars. Since, however, the apparent disc of each star would, at its doubled distance, be diminished in area to one-fourth, so far as regards their combined areas, the increased number of stars would exactly counterbalance the effect of their reduced discs, and the stars upon the second surface would cover the same extent of sky as those upon the first. By the same reasoning this would also be true of those upon the third, fourth, and every succeeding surface; so that, if the stars extend through space without limit, upon including a sufficient number of surfaces the whole face of the sky would at length be entirely covered by stars. Further, as the apparent brightness of an object does not diminish with its distance, unless its light undergoes absorption during its passage from the body to the eye, the whole vault of heaven would be both by day and by night resplendent with a dazzling brilliancy equal to that of the Sun itself—except for a few black dots marking the positions of the interposed planets, and for a group of small dusky objects, spots upon the Sun, by which alone its position in the heavens would be apparent. The actual appearance of the heavens is so strikingly different from this that it is clear that the assumptions lying at the base of the deduction are very wide of the mark.

With reference to this reasoning it must be noticed that the apparent brightness of a surface is independent of its distance, when, and only when, there is no absorption of light in its passage from the surface to the eye. If there is a general absorption of light the argument fails. Less and less light would be received from increasingly distant spheres, and the combined light from all those beyond a certain distance, a distance depending upon the intensity of the absorption, would be a negligible quantity.[1] Absorption of light in interstellar space might result from imperfect transparency of

---

[1] The demonstration of this statement is comparatively simple. Suppose, as a concrete case, that in its passage from the nearest of the spherical surfaces imagined in the argument to the eye, ⅓ of the light that would if there were no absorption reach the eye, is actually absorbed. Then if L represents the whole light that the eye would receive from the stars upon the first surface if there were no absorption, the light actually received from them would be ⅔ L. Consider next the light from the stars upon the second surface. Since the distance separating the surfaces is equal to that from the first surface to the eye; of the light that would reach the first surface from the second on its way to the eye if there were no absorption ⅔ would escape absorption, while in its farther passage to the eye ⅔ of this would escape, the light arriving being therefore ⅔ of ⅔ L, or (⅔)² × L, since L is the same as before, the light received from each sphere upon the assumption of no absorption being the same. By similar reasoning the light received from stars on the third surface will be (⅔)³ L, and so on, the total being

$$\frac{2}{3}L + \left(\frac{2}{3}\right)^2 L + \left(\frac{2}{3}\right)^3 L + \ldots$$

an infinite series of terms, the value of which, by the law of geometrical progression, continually approaches without limit, but can never exceed, 2 L. That is, for this special value of absorption, the whole light from an infinite distribution of stars would be only double the amount that would be received from the stars on the nearest surface if there were no absorption.

the ether of space by which all radiation is transmitted, or from interposed dark matter in the form of dark stars, clouds of cosmic dust, or swarms of meteorites. It is perhaps too much to regard imperfect transparency of the ether as inconceivable; but should absorption in the ether occur, energy in some other form must appear equal in amount to that of the radiation absorbed, and no trace of any such developed energy has been detected. It has, however, been shown in a former chapter that the existence of dark matter in interstellar space is possible if not probable, both in the form of dark stars and as clouds of cosmic dust in the earliest stage of a nebula, while meteor swarms are an obvious fact. Light absorption might arise from all or any one of these causes, cosmic dust clouds appearing the most probably effective, so that the impossibility of an infinite distribution of stars cannot be demonstrated. The distribution is, however, clearly not uniform.

In spite of the serious objections to the view that the density of star distribution upon the face of the heavens can be taken as an indication of the extension of the sidereal system, Sir John Herschel, during his residence at the Cape between 1834 and 1838, extended his father's method of star-gaging, and generally maintained the view that the appearance of the Milky Way is, at any rate in part, due to optical crowding. The comparatively abrupt limits of its stream led him, however, to modify Sir William Herschel's original theory, in assuming the stars to be distributed within the volume

The Milky Way and Star Distribution. 81

of a flat ring, indefinitely extended in all directions, the Solar System being imagined as situated near the centre of the hollow of the ring. By splitting the ring in its medial plane nearly to its centre, and slightly deflecting outward the divided portions, the appearance of the great fissure in the Milky Way was explained. The same fundamental conception underlies the modification suggested by Wilhelm Struve in 1847. According to this scheme the whole of the stars were distributed in parallel layers or strata, the strata being more and more densely crowded towards a central plane very nearly passing through the position of the Sun. According to this view there was a real condensation of stars towards the central plane, which was apparently increased by the effect of optical crowding, the appearance of the Milky Way being the combined effect of the two causes. It was suggested, in addition, that absorption of light in space might reduce the last effect to within narrow limits.

To these modifications of Sir William Herschel's first hypothesis the objections urged against it, though not so entirely overwhelming, are nevertheless fatal. Both are equally inconsistent with the appearance of the Coal Sack and other vacuities, with the dark rifts, and with the minute detail revealed by later researches. According to each of them, the transition from greater to less density of distribution of stars beyond the limits of the Milky Way should be far more gradual and regular than it is. Star-crowding should become less and less pronounced in passing from the limits of the

Milky Way towards its poles; but, in actual observation, while this is found to be the case so far as the average of the stars is concerned, the law is very far from being maintained in isolated regions. Several regions of the sky in close proximity to the poles of the Milky Way are exceptionally rich in stars, while others, closely bordering upon its stream, are among the poorest in the heavens. The views of the two Herschels and Struve are alike untenable, and, since the publication of the last, they have failed to find support save in the pages of certain popular works on astronomy.

The older "disc" and "strata" theories having been found wanting, the Milky Way has come to be generally regarded as a real formation, and attempts have been made to construct in imagination a stream of stars that should give rise to the appearance actually presented. It is obvious that, in its visible aspect, the Milky Way appears as the projection of all of its parts upon the background of the sky. Of the possible depth of the formation the eye takes no cognisance. Its many apparently confused and interlacing wreaths may be, in actual fact, entirely separate, some being projected inward toward the observer, while others may be thrown backward into the more remote spaces behind.

In 1869 Mr. Proctor attempted to show that the chief features of the Milky Way, and more particularly the appearance of the three most pronounced irregularities,—the Coal Sack, the bifurcation, and the great Break in Argo,—might arise from the convolutions of a single stream of stars. One of the

The Milky Way and Star Distribution. 83

diagrams accompanying his suggestion is reproduced, with a slight modification in the arrangement of the lines of vision, in fig. 5. In this, the single stream of the Milky Way is represented as being somewhat fantastically curved, the ends being folded back upon the course of the main stream. The Solar System occupies the position indicated by the point s, and the line of sight s A, escaping into external space between the refolded portions of the stream, accounts for the appearance of the great Break. Close to the Break, in the direction s B, the two portions of the stream are optically superposed, and give rise to the brilliant neck in Argo, while a little farther on one or both portions diverge from the median plane for a short distance, producing the supposed purely optical effect of the Coal Sack in the direction s c. After a farther short distance through which the two portions continue superposed, and through which the stream consequently appears single, apparent division again occurs owing to a second divergence from the median plane, and continues until the branches reunite along s E in the constellation of the Swan. Of these apparent branches, one, the more

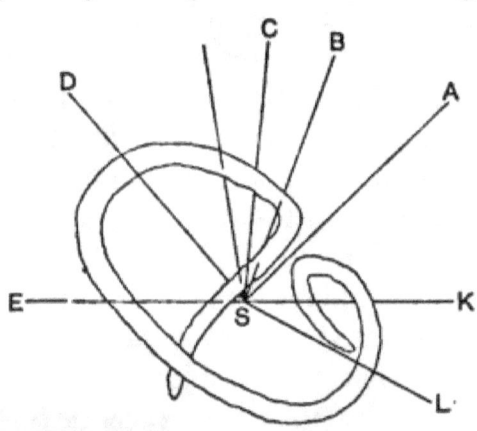

Fig. 5.—Mr. Proctor's suggested explanation of the Appearance of the Milky Way.

northerly, fades and becomes for a time invisible by reason of excessive distance, while the other increases greatly in brilliancy owing to its nearness. From SE to SL the stream appears as it really is, single; but beyond SL, and extending to the great Break, the series of lakes or vacuities, which are a characteristic feature of the Milky Way in this region, are accounted for by repeated temporary deviations of the two portions from the median plane, as in the direction SK.

Mr. Proctor's view of the formation of the Milky Way displays the ingenuity that would have been anticipated from so sagacious and original a thinker, but it is open to objections of so serious a nature as to render its acceptance impossible. It has been urged against it, that, since the brightness of an object does not diminish with increased distance, the fading away and ultimate disappearance of the northern of the two branches bifurcating in the Swan cannot be due to excessive remoteness. It has, however, been seen that the apparent brightness of an object is only independent of its distance if its light undergoes no absorption in space; and since such absorption has been shown to be possible, if not probable, a fatal attack upon Mr. Proctor's scheme cannot be maintained upon this ground. It is curious that Mr. Proctor himself makes no reference to this point, though he was beyond all doubt thoroughly familiar with the elementary law in question. A more serious objection to the suggested explanation lies in the fact that it is scarcely possible to regard the divided branches of the Milky

Way as physically independent. For a considerable distance from the division of the main stream in the Swan, the southern branch continues to cast off incipient streamers and bright projections toward its companion, and the same indication of intimate physical connection between the two is shown in the near neighbourhood of the reunion of the branches in the Centaur. It might seem probable that the varying apparent breadth of the Milky Way might serve as an indication of the proximity or remoteness of different parts of its stream; but it is found that upon the whole the narrowest portions, which would otherwise appear to be the most remote, are the brightest, and it would be indeed difficult to regard enhanced brilliancy as generally associated with increased distance.

The failure of these and other attempts to explain the appearance of the Milky Way, and of the irregularities displayed by it, as being primarily due to perspective effect or optical projection, point to the probability of the more simple and direct view, that the Milky Way is a definite and complicated structure, and that its bifurcation, its vacuities, its gaps, and its other irregularities, are definite physical facts. To this view astronomers have now become reconciled. Adopting it, the sense of the overwhelming mystery of the whole undoubtedly becomes greater, while it must be confessed, that something is gained in the rejection of schemes in which a rather painful suggestion of artificiality somewhat conflicts with the majesty of the problem toward the partial solution of which they have been directed.

The crowding towards the zone of the Milky Way displayed by the naked-eye stars, and more particularly by the brightest of them, suggests that their distribution in space has been controlled either by the Milky Way itself or by the influences to which it owes its being. Not only, however, is a general relation thus indicated between the external stars and the Milky Way, but it has been maintained, notably by Mr. Proctor and Mr. Cowper Ranyard, that there is evidence in many instances of a more direct and intimate relationship between separate stars and groups of stars and the Milky Way. The evidence consists in the arrangement of bright stars, both individually and in groups, with reference to structures in the Milky Way. While acknowledging that in some instances it is difficult to avoid a feeling of doubt that the arrangements in question may not be the results of chance distribution, it is scarcely possible not to recognize in others very strong evidence of intimate physical connection. It must be sufficient here to illustrate the general nature of the arguments adduced by reference to a few selected cases, referring the reader for a more complete discussion of this very intricate and delicate subject to more exhaustive works.[1]

The constellation of the Swan lies in the very heart of the Milky Way, in a region particularly interesting from the evidence of structure apparent in it both to the naked eye and in photographic

---

[1] The subject is developed in considerable detail in the chapter on "The Stars" in *The Old and New Astronomy*, by Proctor and Ranyard; and has formed the subject of a number of articles that have appeared in *Knowledge* during the past fifteen years.

examination. Upon a photographic plate exposed for thirteen hours in the autumn of 1891 by Dr. Wolf of Heidelberg, accumulations of stars are shown of a richness unimaginable in the finest telescopic view, while throughout vast regions the stars appear to be involved in a faintly luminous cloud. This enveloping nebulosity is extensively furrowed by dark lanes and rifts, the borders of which are, in the manner so generally characteristic of them, emphasized by lines of faint stars. Conspicuous in Dr. Wolf's photograph are the images of the bright stars α and γ Cygni, the two that form the head and centre of the familiar cross of the Swan. Upon the photograph, both of these stars appear to be nebulous, their blurred images passing by insensible gradation into the surrounding cloud. It is true that a similar appearance is recognized to some extent in all photographic pictures of bright stars; an extension of the photographic image being caused by the dispersion of light from the point-like and brilliantly illuminated image of the star formed upon the sensitive plate during exposure into the sensitive silver compound around it, but here the want of definition of the images appears to be far more than could possibly be ascribed to this cause, and the unsymmetrical form of the haze by which γ Cygni is enveloped makes it scarcely possible to accept such explanation of the appearance in its case. In addition, the position of α Cygni at the base of a remarkable shrub-like dark formation in the bright nebulosity, and that of γ Cygni as the centre of several diverging nebulous structures, point strongly to a definite physical connection between both stars and the apparently surrounding masses. It appears, therefore, probable that the stars are actually bathed in the depths of the Milky Way in which they appear, and do not owe their appearance in it to the chance effect of optical projection upon it.

The arrangement of many of the stars in the constellation Orion is very remarkable. The probability against three such brilliant stars as those forming the belt falling in a straight line and appearing in such close proximity as the result of chance distribution is overwhelming. The close proximity of the Great Nebula of Orion, as well as the arrangement of many of its contained structures with reference to the direction assumed by the belt, is moreover suggestive; as is also the fact that the belt is situated upon the immediate border of the Milky Way, to which it is very closely parallel. A group of five faint naked-eye stars lies immediately below the belt; they are arranged in a line parallel to it, and it is easy to imagine the line continued upwards and towards the west by a farther stream of five stars separated from the first by a short break and deflected northward towards the Milky Way. A little farther to the west a second stream of naked-eye stars, at least thirteen in number, leaves the constellation in a westward direction, becoming in like manner deflected to the north and actually penetrating the Milky Way. Both of these star streams are represented in a beautiful drawing of the Milky Way as seen by the naked eye, recently executed by Dr. Boeddicker, as involved in its cloudy extensions. To the last point it may, however, be urged with considerable force, that the appearance of a faint nebulous stream connecting the members of a line of faint stars may be an optical illusion, and the failure of the photographic plate to verify the existence of the cloudy stream indicated by Dr. Boeddicker as involving the line of stars immediately below the belt, shows this to have been the case in this instance. From the whole of the evidence it will, however, probably be conceded that the configuration of the stars of the belt, and the symmetrical arrangement of so many conspicuous stars in its neighbourhood in lines having the same general trend, demonstrates the reality of a close physical relation between the whole, while their peculiar relation to the course of the Milky Way renders probable an intimate relation between them and it.

Mr. Proctor has directed special attention to the close association of lucid stars with the course of the Milky Way in the neighbourhood of the Coal Sack. Although by no means devoid of telescopic stars, the dark area of the Coal Sack does not include one visible to the naked eye. If the naked-eye stars had been distributed over the heavens with uniformity, the number assigned to the Coal Sack should have been, at the least estimate, seven; and their avoidance of it is the more remarkable in that in the regions immediately surrounding it they are particularly richly represented. It becomes, therefore, scarcely possible not to regard the bright stars as intimately associated with the Milky Way in this region, and as being situated at nearly the same distance. The brilliant expansion of the Milky Way in which the Coal Sack is situated contains the five bright stars of the Southern Cross; and of these, the most brilliant—the first-magnitude star α Crucis—is actually upon its border. From the appearance of this star in a photograph taken by Mr. H. C. Russell at Sydney in 1890,[1] Mr. Cowper Ranyard has maintained that there is strong evidence of its close association with groups of faint stars that appear upon the plate in its neighbourhood. The great star appears to be the centre of several diverging streams of small ones, while other groups of small stars are arranged concentrically in circles round it.

The apparent luminosity of α Crucis cannot exceed that of the smaller stars by less than three million times, so that if it be regarded as probable that all are equally remote, this proportion is also that of their actual luminosities. No data exist from which it is possible to form an estimate of the distance either of α Crucis or the faint stars, from which it would be possible to compare the actual light-giving powers of the stars with that of the Sun; but if the great star be regarded as being not very different from the Sun in light-giving power, the small ones need scarcely be more than self-luminous planets; while, if the small ones are regarded as the equivalents of the Sun, the large star

---

[1] The photograph is produced in *Knowledge* for June, 1891, and in *The Old and New Astronomy*.

must be of such luminosity that, for its surface brilliancy to be no more than equal to that of the Sun, it must be capable of enclosing within itself the entire orbit of Saturn. Before accepting so astounding a view as the last, it is well to consider whether adequate grounds exist for regarding the former as improbable.

From the accumulation of a considerable body of evidence, the general nature of which has been sketched in the preceding paragraphs, an intimate association appears probable between the naked-eye stars and the system of the Milky Way. The stars immediately below the naked-eye stars in brightness appear to be influenced by its course to a far less degree. It has been seen that the tendency to crowd towards the zone of the Milky Way practically vanishes with stars just below the limit of vision; and any direct relation displayed by these faint stars towards the configuration of the streams of the Milky Way is difficult of detection. Stars of the ninth magnitude are, for instance, distributed with apparent uniformity over the space included between the divided streams between the Swan and the Centaur, scarcely if at all less richly than over the branches themselves; and faint telescopic stars are scattered with approximate evenness over the darkness of the Coal Sack. Thus, in their relation to the Milky Way the naked-eye stars appear to be differentiated from those immediately below them in brightness. Again, since the apparent brilliancy of a star is directly affected by its remoteness, it appears probable that the brightest among the stars are upon the whole those that are nearest. Since individual stars, as has been seen, vary enormously in magnitude, exceptions to such a generalization would of course be expected; they are, in fact, directly illustrated in the near proximity of so inconspicuous an object as 61 Cygni, and in the unfathomable remoteness of Arcturus; but it would be anticipated that, with an increasing number of stars, such irregularities would gradually disappear in a general average. The intimate association of the brighter, and, therefore, in all probability the nearer, stars with the Milky Way, suggests the view that the Milky Way itself is a comparatively near neighbour in space.

By another investigation, proceeding upon entirely independent lines, we are also led to regard it as probable that the brighter of the stars are differentiated from the rest in a special manner. From the time of Ptolemy it has been the custom to classify stars in "magnitudes" in the order of their increasing faintness. To very bright stars, such as Aldebaran and Altair, a position in the first magnitude is assigned; visibility to the naked eye terminates under favourable conditions at about the sixth magnitude; the penetrating power of the largest telescopes probably reaches stars of the fifteenth magnitude, while the power of the photographic plate possibly extends to some four or five magnitudes beyond. Since, however, the measurement of the brightness of a star is a matter of considerable delicacy, and, in fact, has only become satisfactorily possible in recent years, magnitudes assigned to stars by different astronomers have been largely a matter of individual judgment, and it is not surprising that the scales that have been adopted are very conflicting. Since the middle of the present century, however, at the original suggestion of Pogson, it has become the custom to apply to the term magnitude a more definite meaning than it had previously received. It had been noticed by Sir John Herschel, that, according to all generally accepted scales, stars of the first exceeded those of the sixth magnitude in luminosity very nearly 100 times, and Pogson suggested that a ratio in light-giving power of 100 to 1 should be regarded as the definition of a difference of five magnitudes, the four intermediate magnitudes being interposed in such a manner that stars of any one magnitude should bear a constant ratio in their luminosity to those of the magnitude following. To satisfy this condition, it follows that any star must exceed in brightness another of one magnitude fainter by 2·512.. times, this "light-ratio" being the fifth root of 100.[1] In the result, the estimations of magnitude by the older astronomers are found to conform very closely to the absolute scale so far as the naked-eye stars are concerned, but to deviate considerably from it for fainter ones.

From the absolute definition of star magnitude it is possible to calculate the relative numbers in which

---

[1] Thus a first-magnitude star is equivalent to 2·512 second-magnitude stars, a second-magnitude star to 2·512 third-magnitude stars, and so on. Hence a first-magnitude star is equivalent to 2·512 × 2·512 or (2·512)^2 third-magnitude stars, (2·512)^3 fourth-magnitude stars, and (2·512)^5 or 100 sixth-magnitude stars.

stars of different magnitudes should appear in the heavens upon the assumption of their uniform distribution in space. From the result of the calculation, the details of which may be left as an exercise in geometry to the reader and do not present any serious difficulty, it appears that if a number of stars, either of the same or different degrees of intrinsic brilliancy, were scattered in space, subject only to the condition that those of each degree of luminosity were distributed uniformly, there should appear nearly four—more accurately 3.981—times as many stars of any given magnitude as of the next exceeding it in brightness. For every star of the first magnitude there should appear, for instance, nearly four stars of the second, sixteen of the third, and sixty-four of the fourth. It cannot fail to be interesting to compare these numbers with those actually observed.

Data for such a comparison are supplied in star catalogues. Of these the most extensive that has so far been constructed is known as the *Bonn Durchmusterung*. It was compiled under the supervision of Argelander between 1859 and 1862, and in it are recorded the positions and magnitudes of 324,198 stars,—all those of the northern hemisphere down to the 9.5 magnitude. The more recent catalogue constructed at Harvard by Professor E. C. Pickering, giving the magnitudes of stars as measured by the meridian photometer, is undoubtedly the more accurate in this respect, but, since it includes only those brighter than the 6.5 magnitude, it is not so well adapted to the present purpose. The same general result, however, appears from the adoption of the data of either the Bonn or the Harvard catalogues, as well as from Dr. Gould's catalogue of stars visible from the southern hemisphere.

The result of the comparison is expressed in the following table. In the second column are given the numbers of stars between the limits of magnitude indicated in the first. These numbers are given by Littrow as the result of his examination of the *Bonn Durchmusterung*. The third column contains the numbers of stars that should have appeared upon the hypothesis of uniform distribution; and the last, numbers obtained by dividing the figures in the second column by those in the third, that is, the numerical proportion of stars actually observed to those that should have been observed upon the uniform-distribution hypothesis, a quantity that may be conveniently described as the "apparent crowding".

COMPARISON BETWEEN THE OBSERVED NUMBERS OF STARS OF DIFFERENT MAGNITUDES WITH THE THEORETICAL NUMBERS UPON THE HYPOTHESIS OF UNIFORM DISTRIBUTION IN SPACE.

Limiting magnitudes.	Numbers of stars actually observed.	Theoretical numbers.	Apparent crowding.
1 to 2	10	4	2·5
2 to 3	37	15	2·46
3 to 4	130	58	2·24
4 to 5	312	234	1·33
5 to 6	1,001	931	1·08
6 to 7	4,386	3,705	1·18
7 to 8	13,822	14,751	·94
	19,698	19,698	

The general result of the comparison, as expressed in the figures of the last column, is very remarkable. There appear to be many more stars of the first five magnitudes than there should be according to the hypothesis of their uniform distribution in space. Rejecting the stars tabulated between the limits of the first and second magnitudes as being too few from which to draw any reliable deduction, and also as comprising several stars, such as Arcturus and Vega, which should in strictness be excluded, as being too bright to be regarded even as first-magnitude stars, there is apparent in the record a crowding of the brighter stars, that diminishes with increasing faintness and becomes insignificant beyond the limit of the fifth magnitude—that is, near the limit of visibility to the unaided eye. It has been stated that a similar appearance of crowding is recognizable in the more exact record of star magnitudes contained in the Harvard catalogue, as well as in Dr. Gould's catalogue of stars of the southern hemisphere.

An apparent crowding of the brighter stars would be quite consistent with their uniform distribution if light experienced absorption in space, since such absorption would affect the light from distant stars to a greater extent than that from nearer ones. This explanation, however, scarcely appears to be applicable here, since from the fifth magnitude to the eighth, stars appear in numbers not very different from those estimated upon the assumption of their uniform distribution. Were there appreciable light absorption within the limits of space in which stars as far as those of the eighth magnitude are distributed, its effects should be as conspicuous in the falling off of numbers in successive magnitudes among the fainter as in brighter magnitudes.

The most simple view to take with reference to the apparent crowding of the brighter stars is that it results from a real crowding of stars in the neighbourhood of the Sun. There is nothing inherently improbable in this view, since the study of the sky reveals numerous analogous instances of the clustering of stars. Setting aside such extreme cases as are presented in the Pleiades, the Hyades, and other such strongly pronounced clusters, many rich regions of the heavens, not even included in the Milky Way, furnish instances in which the local density of star distribution is far in excess of that which could have resulted from chance distribution.

The suggestion of a clustering of stars in the neighbourhood of the Sun acquires additional interest from the indications already recognized, that the nearer among the stars are differentiated from the rest by an intimate association displayed by them towards the stream of the Milky Way. It appears scarcely possible not to recognize in the complete testimony a suggestion that the Sun is a member of a star-cluster, one in which the Milky Way is involved as a stream of stars far smaller than the more conspicuous members of the cluster, but closely associated with the fundamental scheme of its structure. According to such view the appearance of uniformity in the distribution of the stars from the fifth to the eighth magnitude, as well as their comparative indifference to the zone of the Milky Way, are alike due to their lying for the most part beyond the region in which the clustering tendency and the apparently attractive influence of the Milky Way extends. The Milky Way, together with the cluster containing the Sun, may conceivably constitute a true independent system, while it is possible that similarly associated with other star-clusters there may exist other streams of star-dust, undistinguishable from their excessive remoteness.

Before regarding this speculation as probable, it is essential to imagine some possible explanation of the crowding towards the Milky Way again exhibited by the still fainter telescopic stars. It is not inconceivable that this appearance may be due to the escape of true Milky-Way stars from within its stream into external space. It is not possible here to develop this suggestion fully, but it would appear probable, that, if a number of stars were distributed at random and with random velocities both as regards magnitude and direction through a definite region in space, a condition of things would result not unlike that imagined in the kinetic theory of gases. Pairs of members of the swarm would from time to time approach so closely as to describe, under the influence of their mutual gravitation, hyperbolic orbits round each other, the common centre of mass of the pair marking the position of a common focus. If, by chance, it happened that the masses and velocities of the pair were so related that their centre of mass was at rest, the direction only of the star motions would be affected by their near approach, the velocity of each being reduced upon separation to an extent equal to its increase upon approach; but if, as would generally be the case, the centre of mass was in motion, since the velocities of the stars would be ultimately unaffected with reference to it, the actual velocities would be changed, the star receding after the "encounter" in the direction of motion of the centre of mass having its velocity increased, while the speed of the other would have become less. Encounters between stars continually occurring, all velocities, without limit of magnitude, would be continually being produced in the cluster, and from time to time a star would acquire sufficient speed to carry it beyond the limits of the cluster, while its speed might be so great as to place it beyond the controlling influence of gravitation, in which case it would leave the cluster never to return. The system would slowly disintegrate, and during the process the escaped members would be found scattered in external space most densely distributed in the immediate neighbourhood of the original swarm.

Similar encounters must occasionally take place between members of the main cluster in which the Milky Way is involved, which, by similar reasoning, it is scarcely possible to regard as a stable system. The extreme velocity of such "runaway" stars as 1830 Groombridge may well be due to their having experienced a number of favourable encounters with other stars, and in any case does not point to their being, as has been suggested, temporary visitors to the system of the visible stars from external regions of space, ploughing their way through it by reason of enormous initial speeds incapable of generation by the gravitational attraction of the system. It is conceivable that in the remote past the sun-cluster may have been far richer than it is now, and the firmament may have been more resplendent with brilliant stars, but that from age to age its members may have been gradually scattered, and the vault of heaven may now be growing poorer.

It is scarcely necessary to remind the reader that in this chapter no attempt has been made to explain the function of the Milky Way, or its connection with the stars. An attempt only has been made in the latter pages to define its possible relation to the stars, and it is not suggested that the attempt extends beyond the limits of speculation. Ignorance of the distances of more than an insignificant minority of its members appears at present an insuperable obstacle towards extending the web of exact knowledge far into the system of the stars; but, in the absence of more exact methods, it is impossible for the lover of the picture of infinite grandeur and majesty, mapped out night by night upon the fair face of the starlit sky, to refrain from indulging in some conjecture, however vague and in itself unsatisfactory, as to the meaning of so exquisitely beautiful and mysterious a record.

Chapter III.

The Recent Study of Mars.

From the time that the telescope revealed to Sir William Herschel the first clear picture of the planet Mars, and led him to regard the details of the delicately-tinted image presented to his view as indicating the existence upon the surface of the planet of physical conditions not very unlike those familiar to the inhabitants of the Earth, a special interest has attached to this, the only one of the orbs of heaven that it is possible to contemplate with any degree of confidence as a sister world. In their distribution, general configuration, and colour, the planetary markings have appeared during the greater part of the present century to harmonize well with their tempting interpretation as oceans, continents, and polar regions bound in eternal snow. The rotation of the planet and the haze in which many of its features appeared to be enveloped indicate the regular succession of day and night, each passing into the other by the insensible gradations of morning and evening twilight; while the tilt of the axis of rotation of the planet to the plane of its orbit demonstrates the constant recurrence of seasons. In recognizing upon a planet so many conditions essential to its well-being as a world, it has been impossible to restrain imagination from supplementing actual discovery in regions lying beyond the power of telescopic observation.

The lands and waters of Mars teemed with animal and vegetable life. In lands over which the cold and heat of winter and summer and day and night never ceased, seed-time and harvest were added; while, passing their brief span of struggle and passion, and fighting to maintain their mastery over nature, were intelligent beings, towards whom, in imagination, the right hand of fellowship was longingly extended across a separating chasm of nearly 50,000,000 miles.

For the greater part of the century that followed Herschel's observations, although knowledge of Martian detail steadily increased, little was added to it materially to affect the nature of the picture drawn by him. In the drawings of Beer and Mädler, Dawes, and other astronomers, as in the exquisite pictures constructed by Green from his study of the planet from Madeira during its specially favourable appearance in 1877, the planet, though shown in greater detail and perfection, was essentially the Mars of Sir William Herschel, and his view of Mars as "a miniature of the Earth" appeared to derive additional confirmation. More recently, however, interest in Mars has been reawakened and maintained at the highest pitch by the alleged appearance upon the surface of the planet of a variety of detail of the most unexpected and perplexing kind. The "canal system" of Mars, the first suspicion of which was suggested to Schiaparelli in 1877, if it has led to speculations that scarcely add to the dignity of science, has renewed the youth of Martian study, and has directed towards the

The Recent Study of Mars. 103

planet a keen scrutiny only rendered possible by the construction of the great telescopes of modern times. In the present chapter an attempt will be made to trace the course of recent discovery upon Mars, and to discuss, though necessarily imperfectly, certain views that have been suggested, and some difficulties that have arisen, in the attempted interpretation of the appearances.

The planet Mars lies next the Earth in order of increasing distance from the Sun, the distance of Mars being 141,500,000 miles, while that of the Earth is rather less than 93,000,000. The respective distances are therefore in the proportion of 1·523 to 1, or nearly of 3 to 2, a relation that will be useful in the sequel. The diameter of Mars is 4230 miles, that of the Earth being 7918 miles, from which it follows, from simple geometry, that in volume the Earth exceeds Mars by 6·57 times. From disturbances produced by Mars in the movements of other members of the Solar System, by its gravitational attraction upon them, it appears that its mass—or quantity of contained matter—is less than that of the Earth in the proportion of 1 to 9·34. Consequently, the Earth being 9·34 times as massive while only 6·57 times as bulky as Mars, the density—or mass of a given bulk—of the Earth must exceed that of Mars in the proportion of 9·34 to 6·57, or of 1·42 to 1. It also follows from these data that the intensity of gravitation exercised by Mars upon bodies at its surface must be less than that exercised by the Earth upon bodies at its surface in the proportion very nearly of 2 to 5, that is, a body the weight of which had been determined at the surface of the Earth, would, if it were transferred to the surface of Mars, weigh only two-fifths as much. Like the Earth, Mars travels round the Sun in an orbit that is, although nearly circular, slightly elliptical, the planes of the two orbits very nearly coinciding. The period occupied by the planet in completing its orbit—that is, the year of the planet—is 686·9 days, the Earth's year being 365·26 days. The longer period of Mars is due, partly to the greater length of its orbit, and partly to the planet's speed in its orbit being less than that of the Earth in its orbit, a consequence of its greater distance from the Sun.

Like all planets the orbits of which inclose that of the Earth, Mars is seen to best advantage when in "opposition" to the Sun—that is, at the instant at which the Earth, overtaking the planet in its slower journey, passes directly between it and the Sun. Under these conditions, not only is the distance separating the Earth from Mars less, and the apparent size of the planet therefore greater, than at other times; but the hemisphere of the planet that is illuminated by the Sun is presented directly towards the Earth, so that the disc appears "full". At other times, when the illuminated hemisphere is not directly presented to the Earth, the planet exhibits phases resembling those of the Moon when not far from the full. The phases of Mars show that, like the Earth and Moon, it is not inherently luminous, but that it is rendered visible by sunlight scattered from its surface. It is clear that, as is the case with the Moon when full, a planet in opposition to the Sun must rise in the east as the Sun sets in the west, and, after ascending the heavens during the evening hours and attaining its greatest altitude in the south at midnight, must descend towards the west in the early morning, setting at sunrise. Hence, a further advantage of a planet's being in opposition arises from its being then visible through all the hours of the night.

If the orbits of the Earth and Mars were circles lying in the same plane and having the same centre, and if the Sun occupied the common centre, then, at every opposition, no matter what position in its orbit the Earth might happen to occupy, its distance from Mars would be the same. Under such circumstances Mars would appear under the same conditions at every opposition, and all would therefore be equally favourable. These simple conditions do not, however, exist. The planes of the orbits are indeed so nearly coincident that their deviation from perfect coincidence may for the present purpose be ignored; but the forms of both orbits are ellipses, deviating, especially in the case of the orbit of Mars, appreciably from the circular form; while, in accordance with Kepler's first law of planetary motion, the Sun is situated, not in the centre, but in a focus common to each ellipse. The orbits of the Earth and Mars, and the Sun's position relatively to them, are represented to scale in the accompanying figure (fig. 6), and it is interesting to notice that the ellipticity of each orbit is indicated far more clearly in the displacement of the Sun from the centre than by deviation from circularity in outline, which indeed, even in the case of the more elliptical orbit of Mars, is probably inappreciable to the most critical eye.

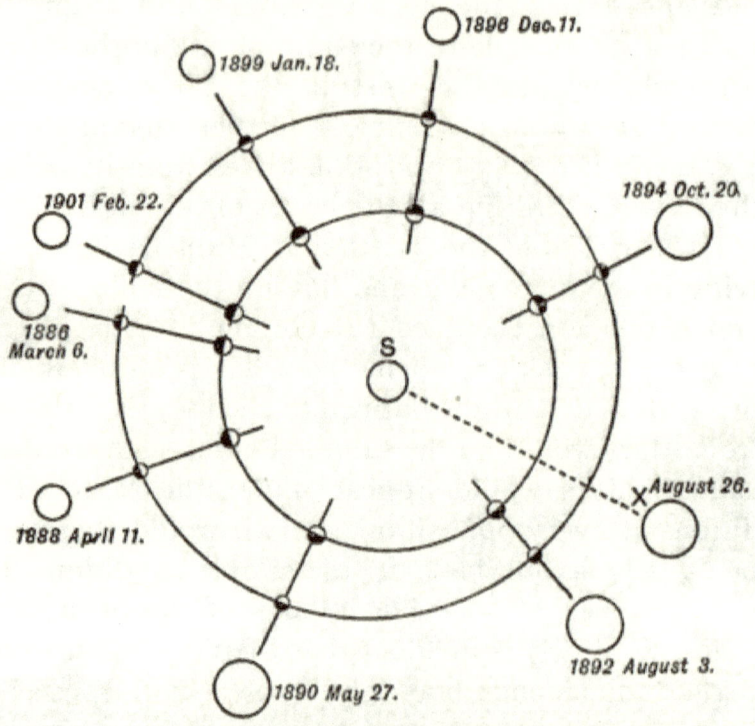

Fig. 6.—Oppositions of Mars.

From the eccentricities of the orbits of the Earth and Mars, and from the position of the Sun relatively to them, it follows that, in one direction—indicated by the line sx in the figure—the distance between the orbits, as measured along a straight line radiating from the Sun, is least. If, therefore, at the time that the Earth is crossing this line, Mars also happens to lie upon it, an opposition will result which will be the most favourable possible; the planet then appearing brighter to the naked eye, and presenting a larger disc when viewed through the telescope, than at any other time; while other oppositions will be more or less favourable according as to whether the direction of the Earth and Mars as viewed from the Sun is nearer or farther from the line of most favourable opposition sx. The Earth in its annual journey round the Sun crosses the line sx upon the 26th of August in each year; hence, the nearer to this date of occurrence, the more favourable is an opposition of Mars.

Since the periods of revolution of the Earth and Mars round the Sun are 365·26 and 686·9 days respectively, it follows, from simple arithmetic, that, upon the average, the Earth must overtake Mars, and an opposition must therefore occur, at intervals of 780 days, or nearly two years and two months. This would be the constant interval between any two successive oppositions if the orbits were circles with the Sun in their common centre, and if, as would then necessarily be the case, the speed of each planet were uniform. Owing, however, to the elliptical forms of the orbits, and to the fact, expressed in Kepler's second law of planetary motion, that the velocities of both the Earth and planet vary with their distance from the Sun, the intervals between successive oppositions are sometimes greater and at other times less than the average, the exact calculation for particular cases being a very laborious matter. In the figure the positions of the Earth and Mars are given for all oppositions occurring between 1886 and 1901. It will be seen that oppositions are now (1898) becoming less and less favourable, and that they will continue to deteriorate until 1901, in which year an opposition will occur under almost the most unfavourable conditions possible. After 1901, however, improvement will set in, culminating in fine oppositions in 1907 and 1909. The circular discs arranged outside the orbit of Mars in the figure represent to scale the relative apparent sizes of the disc of Mars as seen from the Earth at the different oppositions. They show in a striking manner the special advantages attending oppositions that occur in the early autumn months.

Viewed through a fine telescope and under favourable conditions, Mars, when in opposition, presents a picture of singular beauty and charm. Markings, some so distinct as to be clearly recognized at a first glance, others less strongly pronounced, and others again so faint as to tax the powers of the keenest vision assisted by the finest optical power, are distributed over the disc-like picture; while the beauty of the spectacle is enhanced by the presence and variety of colour, and by exquisite gradations of tint in different regions.

Upon continuing the study of the planet, it soon becomes evident that change is in progress, not in the form of the features themselves, but in their positions relatively to the outline of the planet. Details first seen near the centre of the disc have drifted to the left; others, originally near the left limb, have disappeared; while others, previously invisible, have appeared within the right-hand limb. These changes clearly indicate, that, like the Earth, Mars is in rotation. Further, it becomes apparent that the period of rotation of the planet does not differ very much from that of the Earth, for in little more than twenty-four hours the picture presented is again that originally seen.[1] Day and night, the appearance of diurnal revolution of the heavens, as well as all other celestial phenomena resulting from the rotation of a planet, follow therefore upon Mars with the same regularity, and at nearly the same rate, as upon the Earth.

From the study of the planet's rotation it is a simple matter to determine the position within it of the axis about which the rotation takes place. When this is done, it is found, that, as is the case with the Earth, the axis of rotation of the planet is inclined to the plane of its orbit round the Sun, the inclination of the axis (24° 50') curiously approximating in value to that of the inclination of the Earth's axis to the plane of its orbit (23° 27'). The tilt of the axis of rotation gives rise to the phenomena of seasons; hence upon Mars, spring, summer, autumn, and winter follow with the unceasing regularity familiar to inhabitants of the Earth.

A further point of similarity between the physical conditions existing upon Mars and those upon the Earth becomes apparent from the study of the planet's rotation. As different features are carried by the rotation towards the left-hand limb, they disappear while still at an appreciable distance from

[1] The period of rotation of Mars is 24 hours 37 minutes 23 seconds.

it, melting into a luminous ring known as the "limb-light", that appears to continually cling to the outline of the planet, extending inwards for some distance from the limb. In a similar manner, features brought into view by rotation do not at once appear as they are brought on to the disc, but as gradually emerge from the limb-light upon their side of the planet, only becoming distinctly visible when the rotation has carried them a considerable distance on to the disc.

The suggestion that Mars is enveloped in an atmosphere similar in its physical properties to that of the Earth offers so simple and sufficient an explanation of the limb-light that it is scarcely possible not to regard it as the true one, more especially as the existence of an atmosphere upon Mars is independently demonstrated from the nature of changes continually in progress in the visible features of the planet, to which attention will shortly be directed. According to this view, the appearance of the limb-light results from the scattering of the Sun's rays in the atmosphere of Mars. Simple considerations, such as, for instance, the darkened tint of the sky when viewed from high altitudes, indicate that the appearance of the sky as a vault of deep-blue eternally extended overhead is due to a scattering of the Sun's rays in the atmosphere of the Earth. Tyndall has shown by a series of experiments of extreme beauty that this scattering is in all probability effected by innumerable minute vesicles of water floating in the atmosphere, and that the formation of these vesicles is assisted by, or may indeed be entirely dependent upon, the presence of specks of dust, which form nuclei around which condensation of the vapour of water present in the atmosphere takes place. It is in harmony alike with theory and experiment that the more refrangible of the Sun's rays should experience such scattering to a far greater degree than those less refrangible; so that, of the component colours of a ray of sunlight penetrating a column of atmosphere, the more refrangible colours—those near the violet end of the spectrum—should be scattered in all directions around, while the less refrangible—the red and adjacent rays of the spectrum—should pass through more readily, a law illustrated in the familiar fact of the great penetrative power of a red light in a fog, and also supplying an explanation of the transparent blue of the noonday sky and the crimson colours of sunset. Assuming the existence upon Mars of an atmosphere possessing similar dispersive powers, the limb-light is simply and naturally explained. Bathed in the Sun's rays and containing floating matter capable of scattering them, the atmosphere of Mars would form a luminous shell enshrouding the visible hemisphere of the planet. The line of vision from the Earth directed to the centre of the disc pierces this shell perpendicularly; the portion of its length included in the shell is therefore the least possible, and the illumination of the atmosphere is barely appreciable. The line of sight to the limb, however, meets the visible hemisphere tangentially, and, traversing the air-shell very obliquely, its intercepted length is great, and the illumination of the air very apparent. From the edge towards the centre of the disc the length of the line of vision involved in the atmosphere of the planet continually decreases, the appearance of illumination therefore becomes less and less, and the limb-light is the result.

The fading of the planetary features upon approaching the limb is further aided, first, by the fact that as they approach the edge of the visible hemisphere their actual illumination becomes less, as does the terrestrial landscape towards sunset, both from the increasing slant of the Sun's rays and by the greater absorption exercised upon the rays from the greater length of their atmospheric path;[1] and, secondly, from the greater length of the Martian atmosphere through which they are viewed, and the consequent increased absorption exercised upon the rays in retraversing the atmosphere, after reflection from the surface of the planet.

Explanations other than the one advanced here have been suggested to explain the limb-light. The appearance has been ascribed to the deposition of hoar-frost, upon the approach of night, over regions about to enter the dark hemisphere of the planet; and to the lingering of the frost in the early morning over those that have recently emerged from it. This suggestion appears, however, to be disproved by the observed symmetry of the limb-light in the cold polar and warmer equatorial regions of the planet; and by the fact that, in

[1] This only applies to Mars when in or near opposition.

observations made at times when Mars is not in opposition, and when, consequently, its disc does not appear to be full, the limb-light has been seen to cling to the limb itself in preference to those regions in which morning and evening are indicated by proximity to the terminating line separating the dark from the bright hemisphere.

The tenuity of the veil spread over the illuminated hemisphere by the atmosphere is generally taken to indicate that the surface density of the atmosphere —that is, the quantity of air accumulated over each square mile of surface—is less in the case of Mars than in that of the Earth. It is probable, that if the surface density of the Martian atmosphere were equal to that of the Earth, its veiling effects would be far more pronounced than they are; and that, even in the centre of the disc, the surface markings would be permanently concealed beneath a brilliant haze. Recognizing that the appearance of the sky is the result of the scattering of sunlight in the Earth's atmosphere, it will be apparent that, to an observer who should ascend above the highest reaches of the air, the Earth would appear upon a clear day to be veiled by the blue haze of the sky, now lying between him and the landscape beneath. From actual measurements of the brightness of the sky carried out by Langley it has been concluded that to such an observer all terrestrial features except the most brilliant would be scarcely visible, their fainter light being overwhelmed by the more intense glare of the intervening atmosphere. To a possible inhabitant of another planet, provided with adequate instrumental means, the Earth would appear as a dazzlingly brilliant, but probably a nearly uniformly illuminated orb. The ice-bound regions near the poles and the snow-clad summit of a mountain, might, here and there, be clearly distinguishable in the general luminosity of its disc, but it would scarcely be possible to trace upon it any of the more familiar terrestrial features. Upon Mars, however, surface markings are clearly recognized unless fairly close to the limb, while those in the centre of the disc appear to experience but slightly the effects of atmospheric veiling. It is therefore commonly assumed that in the density of surface distribution of atmosphere, Mars is poorer than the planet Earth.

It must be acknowledged that this reasoning, though lending a strong probability to the view, is not quite conclusive. It essentially rests upon the assumption that the power of an atmosphere to scatter light may be taken as a measure of its density. It has been seen, however, that the scattering of light is effected by solid and liquid matter suspended in the air, and is not, therefore, an inherent property of an atmosphere itself. Were there no floating matter in the Earth's atmosphere, there would be no scattering of light within it. The blue sky, even in the immediate neighbourhood of the noonday sun, would under such conditions be replaced by a vault of intense black, in which, by day as by night, the stars would shine with a lustre unknown even on the clearest and darkest nights. Absorption of light would still occur, but to so slight an extent—such is the transparency of pure air—as to be barely appreciable; while it is hardly necessary to state that the absorbed light would be entirely extinguished, and that no appearance of a sky could result from it. That the atmosphere of Mars contains floating matter in proportion to its density may be true, but it is an assumption that it is not possible to verify. It will be seen later that there are indications of a scarcity of water on the surface of Mars, and that there is a very strong probability that its atmosphere is charged with the vapour of water to a far less extent than is the atmosphere of the Earth. It is probable that the condensation of the vapour of water plays an important part in the dispersive action of the atmosphere on light, and that, therefore, under conditions otherwise similar, a less moist atmosphere would possess a feebler scattering power than another. That Mars possesses a more tenuous atmosphere than that of the Earth may be probable, but an equal or even a greater density is not inconsistent with the telescopic aspect of the planet.[1]

Spectroscopic evidence bearing upon the question of the Martian atmosphere is so curiously conflicting that it is perhaps better to wait for further observations before taking it seriously into account.

Of the different features apparent upon the disc of Mars, generally the most conspicuous are two white patches, nearly always visible in the neighbourhood

---

[1] The *a priori* arguments, based upon the relative volumes and masses of Mars and the Earth, that are frequently adduced as evidence for a rare atmosphere on the planet appear to possess little if any value.

of the poles, though not arranged symmetrically round them. So brilliant are they, that they have been seen sparkling like twin stars at times when the sky has been covered by haze to such an extent that the outlines of the planet itself have been invisible. From their general appearance, as well as from their situation in the immediate neighbourhood of the poles, they have been regarded as accumulations of snow and ice, similar in their nature and in their mode of formation to the polar caps of the Earth. This conclusion is strongly supported by the nature of the changes apparent in both of them during the progress of the Martian seasons. Upon the approach and during the continuance of winter in either hemisphere of Mars, as, in the orbital movement of the planet, the hemisphere is turned from the Sun, the white cap surrounding its pole continually increases, its boundaries extending farther and farther towards the equator; while later, during spring and summer, as the hemisphere is again turned towards the Sun, its white covering dwindles in dimension, becoming generally reduced to an insignificant oval patch. Upon a recent occasion, indeed, when the planet was near its opposition in 1894, the south polar cap entirely vanished, the substance composing it having been apparently dissipated beneath the rays of the summer sun. Other appearances, more rarely recognized in the caps and in their immediate neighbourhood, lend additional support to this view. On June 8th, 1894, Mr. Lowell, while observing Mars from Arizona, saw two points of light of dazzling brilliancy flash out in the midst of the south polar cap. For a few moments they sparkled in the surrounding whiteness and then disappeared. It is difficult to resist Mr. Lowell's interpretation that their appearance was due to the glint of ice-slopes flashing the sunlight towards the Earth, as, during the rotation of the planet, the slopes were for a few moments placed at the proper angle to the rays. Similar appearances had been noticed by Mr. Green during the opposition of 1877, but in this case they were seen near to, but not actually involved in, the polar cap.

Distributed round the planet in a rough zone appreciably parallel to its equator, and extending over considerably more than a half of its entire surface, are a number of patches, generally of a soft rounded outline, and of a colour that has suggested the orange-yellow of a field of ripe corn. It is to these that the planet owes the familiar ruddy tint that has caused it to be associated in name with the god of war. Bounding, and frequently deeply indenting, these orange masses are regions of a gray-green tint, and these complete the picture of the planet's surface as seen with moderate optical power. Encouraged by the close similarity in appearance and behaviour between the polar caps of Mars and the Earth, it has been the custom to pursue the analogy further, and to see in the orange patches and in the gray-green markings upon Mars the continents and oceans of a miniature world.

That the orange masses upon Mars are indeed land appears probable, from the similarity in their appearance to that which it may be well supposed the great deserts of the Earth would present under similar conditions of observation, as well as from the permanent appearance upon them of delicate markings revealed by higher optical power; though, perhaps, as strong an argument as any lies in the difficulty of suggesting any other explanation for their appearance. That the gray-green markings are the surfaces of Martian seas appears at a first glance a scarcely less plausible suggestion. Their colour is not unlike that of water; the existence of extensive tracts of water upon Mars harmonizes well with the view of the polar caps as accumulations of snow and ice; and their aqueous character was supposed to have received its final confirmation in 1867, from the announcement of Sir William Huggins, that, from the spectroscopic examination of the light from Mars, he had detected the existence of the vapour of water as a constituent of its atmosphere. Of late years, however, considerable doubt has been thrown upon this rather attractive view. In 1877 Schiaparelli of Milan maintained that, if the gray-green markings were the surfaces of water, they should occasionally, when turned at the proper angle to the directions of the Sun and Earth, unless indeed their surfaces were continually in a state of violent disturbance, reflect the Sun's rays in such a manner that its image should appear as a bright star sparkling upon them. It is not difficult, upon the assumption that the water surface is clean and still, to calculate the intensity of the solar image that should be formed under the actual conditions, and it appears that it should be so brilliant as to be readily capable of recognition. No such appearance has, however, ever been recorded upon the disc of Mars.

A very interesting though not in itself a conclusive observation has recently been made by Professor W. H. Pickering, in the examination of Mars under the polariscope. It is well known that the fraction of light that is regularly reflected from the surface of any transparent substance exhibits the phenomena of polarization—it is capable of being again reflected more or less effectively by a second transparent surface, according to its direction of incidence upon it; it is transmitted through certain crystals—such as tourmaline—more or less readily, according to the direction of the axis of the crystal; and upon traversing many crystals, and in its subsequent analysis by a second polarizing apparatus, it is capable of developing the exquisite effects of colour familiar to many observers with the microscope. It is a simple matter to detect polarization in light reflected from a plane glass or a water surface, especially for certain angles of incidence, and in the light of the sky, which is strongly polarized as the result of its scattering by water vesicles suspended in the atmosphere. In 1894 Pickering examined the light from the gray-green markings upon Mars with a specially-constructed polariscope, but failed to detect in any of them any trace of polarization.

There is no doubt that if polarization had been evident in the light of the gray-green markings, their liquid nature would have been demonstrated;

but the interpretation of the negative evidence is not so definite. Polarization would be produced by regular reflection—by which is meant reflection as from a mirror, the angle of incidence being equal to that of reflection—or it might conceivably result from the scattering of light by fine particles suspended in a body of water, in a manner analogous to that by which the appearance of the sky is produced. If, however, the surface of water were not clean, the impurities upon it would scatter light incident upon it in all directions, and in light so scattered no polarization should be apparent; there would, in fact, be no water surface exposed to the light, but a dirt surface concealing a water surface beneath. Pickering's observations would appear to indicate that if the aqueous view of the gray-green markings is to be retained, it must be modified in this direction. The same modification would also account for the non-appearance of the image of the Sun upon the surfaces of Martian seas.

Still more recent and direct observations appear to involve a complete refutation of the aqueous character of the gray-green markings on Mars. Mr. Douglass at Arizona in 1894, and Mr. Barnard at the Lick Observatory in 1896, have succeeded in distinguishing over the entire surfaces of them a considerable amount of delicate and permanent detail, an intricate tracery clearly inconsistent with the older view as to their nature. Mr. Barnard, in particular, examining the planet with the superb refractor of the Lick Observatory upon Mount Hamilton, under atmospheric conditions that frequently approximated to perfection, describes the detail revealed in the regions of the so-called seas as being so intricate, small, and abundant, that it baffled all attempts to properly delineate it. He suggests that, to those who have looked down upon a mountainous country from a considerable elevation, some conception of the appearance presented may be formed. From the appearance of the country round Mount Hamilton as seen from the observatory, it was possible to imagine that, as viewed from a great altitude, this region, broken by cañon, slope, and ridge, would closely resemble the surface of the Martian seas. During the observations the conviction seemed to force itself upon the observer that he was actually looking down from a great elevation upon just such a surface as that above which the observatory was situated.

It appears, therefore, that if water exists at all upon Mars in the liquid form, it must be sought elsewhere than in the so-called seas; and it is possible that, in an observation made in 1894 by Mr. Lowell and Professor Pickering, its place upon the surface of the planet was revealed for the first time. During a careful study of Mars, when near its opposition in that year, with the aid of a fine refracting telescope of 18 inches of aperture, there appeared a dark belt forming a fringe to the south polar cap. The belt first appeared after rather more than a Martian month following the spring equinox of the planet. It was estimated as being the darkest marking on the disc, and appeared to be of a decidedly blue colour. As the polar cap dwindled, the belt followed, clinging to its edge. At midsummer upon Mars it was described as a barely discernible thread drawn round the minute white patch, which was all that then remained of the enormous snow-fields of some months before. Finally, when the cap vanished, the spot, where its girdle, long since too small for detection, had existed, had become one yellow stretch.[1]

That the belt seen upon this occasion was water, or at any rate liquid formed by the melting of the polar cap, appears a plausible suggestion, and appears more probable from the fact that Professor Pickering, on subjecting it to examination with the polariscope, was convinced that its light showed marked evidence of polarization. The interpretation of the sequence of the observed phenomena appears to be—that the melting of the polar cap gave rise to a fringing belt of liquid, which first appeared as such, but was rapidly distributed over the summer hemisphere in streams too fine for detection.

A dark belt surrounding the north polar cap had been seen as early as 1830 by Beer and Mädler, and other like appearances, which may have been of the same nature, have been recorded by other astronomers.

The atmosphere of Mars appears to be in striking contrast with that of the Earth, in its almost entire freedom from cloud or mist. From time to time during the study of the planet, extensive regions have appeared, sometimes for a considerable time,

[1] *Mars*, by Percival Lowell.

to be indistinct, the result, it has been generally supposed, of accumulated cloud or mist; but as more perfect optical means have been applied, and as observations have been conducted from localities specially selected for their atmospheric steadiness and the consequent improvement in the definition of the telescopic picture the appearances have become less and less frequent. During the entire course of a series of observations upon the planet, continually maintained at Flagstaff in Arizona, under the direction of Mr. Lowell, from May to the end of November, in the year 1894, no case of obscuration that could be ascribed to cloud or mist was recorded by anyone of the three astronomers engaged in the work, with the exception, perhaps, of some minute white specks, limited in position to the immediate neighbourhood of the line of division between the bright and dark hemispheres, and which may have been transient morning and evening clouds. During the actual progress of these observations, however, other astronomers, observing Mars with less perfect optical means and under less favoured atmospheric conditions, believed that they recognized one of the most extensive formations of cloud that has ever been recorded. To Mr. Stanley Williams at Brighton, for instance, the greater part of the Miraldi Sea, one of the largest, darkest, most definite, and most characteristic of the green regions, disappeared almost entirely from view, apparently densely obscured by cloud or mist.[1] There is no doubt that changes in the tint of several

[1] Observatory, 1894, p. 391.

regions of the planet's surface are of frequent occurrence, and it is possible that such changes, which may cause the disappearance of detail, especially if accompanied by a general lightening of tint, may have been interpreted as cloud and mist. In the apparent absence of cloud, and, consequently, of rain, upon the surface of the planet, it is probable that the polar caps are formed by the continued deposition, as hoar-frost, during the long Martian winter, of the vapour of water or of some other liquid present in the atmosphere.

When the planet was near its very favourable opposition in 1877, Schiaparelli at Milan, while observing with a telescope of rather over 8 inches in aperture, detected certain faint dusky lines projecting from the gray-green regions well into the interior of the orange continents. The streaks appeared to be most conspicuously visible shortly after the mid-winter of the hemisphere in which they appeared. At the rather less favourable opposition of 1879, the streaks first seen were traced, accompanied by others; to Schiaparelli they appeared to be more sharply defined than before; while one of them appeared to be double, consisting of a pair of parallel streaks separated by a distance of between one and two hundred miles. As before, as well as at succeeding oppositions, the streaks, to which Schiaparelli had now given the most unfortunate name of "canals", appeared more clearly during the latter part of the Martian winter and the early spring. At succeeding and increasingly unfavourable oppositions, the numbers, length, and instances of duplication, of the canals were steadily increased; in character they seemed to be more rectilinear and more sharply defined; and, to Schiaparelli, they at length appeared to form a reticulated network, extended over nearly the whole of the orange continents. Until 1896 to no other observer had the canals so much as appeared, but in that year a few were recognized by Perrotin and Thollon at Nice, by the aid of a then newly-constructed telescope of 29 inches of aperture. As oppositions again became more favourable, however, they appeared to quite a number of astronomers supplied with the most ordinary instrumental means; and they have now become entirely notorious.

According to the evidence of astronomers to whom they have appeared, the canals are faint lines that appear to become finer and straighter as the eye becomes accustomed to their appearance. Their width is estimated as being not less than fifteen, or more than sixty, miles. They follow, as closely as can be seen, the course of great circles upon the surface of the planet,[1] and can be frequently traced for upwards of 1500 miles. They mutually intersect in a most remarkable manner, several of them frequently passing through the same point, from which, again, it is not uncommon to find other canals originating, so that the entire

---

[1] A great circle of a sphere is the circle that divides it into two equal parts. It is the largest circle that can be drawn upon the surface, and its course marks the shortest line that can be drawn between two points on the surface. Great circles are illustrated in meridians of longitude. Parallels of latitude are known as small circles.

surface of the planet appears as if involved in a complicated network of delicate tracery. Before the year 1894, canals had only been recognized upon the orange continents; but as the planet approached opposition in that year they appeared to Mr. Douglass at Flagstaff to be distributed scarcely less richly over the green of the so-called seas. Points of intersection of canals are frequently emphasized by the occurrence at them of round or oval dusky spots, which have received the name of "lakes". The system of canals and their associated lakes varies in visibility with the Martian seasons, being commonly invisible during winter, gradually appearing in the early spring, and again disappearing during the progress of late summer and autumn.

It has been suggested that the canals of Mars are waterways, and that their emergence from invisibility upon the approach of spring may be due, either to the dissipation of their winter covering of ice and snow, or from their becoming extensively flooded by large volumes of water discharged into them by the melting of the polar snows. Their light revealed no trace of polarization when examined by Professor W. H. Pickering, from which he has advanced the conjecture that the streaks may be tracts of country lying upon either side of water-courses, themselves too fine for detection, and irrigated by them, rather than water-courses themselves; and that their appearance in the spring may be due to a general growth of early vegetation over them as they become fertilized by the flooding of the streams.

The apparent regularity of the canals, as well as the difficulty of suggesting any other explanation for them, have been at times regarded as indicating artificiality. According to Mr. Lowell, who is a strong advocate of this view, the canals and their connected lakes, which, according to this view, may be more suitably regarded as oases, are the visible result of an extensive system of irrigation carried out by intelligent beings on Mars. For the inhabitants of Mars, as for man, water is a necessity of life; and since water appears to be scarce on the planet, being, indeed, apparently only to be obtained from the melting of the polar snows, the inhabitants have, with consummate engineering skill, constructed an extensive network of channels, extending from the polar regions over the entire surface of the planet. By these channels, upon the melting of the polar snow, the lower lands are well supplied with water, the vegetation springing up on them giving rise to the appearance of the gray-green tracts; while the irrigation of the higher and desert districts is confined to the immediate neighbourhood of the channels, and results in growth of vegetation over belts of country irrigated by them on either side, and the oases at their junctions. From these follow the appearances of the "canals" and "lakes".

So much for the outlines of a romance, the leading features of which have become, largely through the co-operation of "our own correspondent" and Mr. Lowell, familiar to the greater number of readers of the daily press about the times of recent oppositions of Mars. To those not practically acquainted with the extreme delicacy involved in the telescopic observation of detail so faint as just to hover upon the verge of the visible and the unseeable, it must appear that, to the concurrent testimony of so many laborious observations, conducted by astronomers, some of established reputation, and the greater number of unquestioned honesty of purpose, there can be but one interpretation; and that, upon the surface of Mars, features and a system alike unique in the revelations of the Universe have been firmly established. The examination of more complete evidence, however, suggests grave objections to the unhesitating acceptance of this view, while to many thoughtful observers it has appeared hard to escape from the conclusion that the complicated meshes of the canal system upon Mars must be regarded as little more than optical illusions, faulty interpretations of the faintest shades of tint, the exact nature of which has not so far been established.

Although originally discovered, and the courses of many of them traced, by the aid of a telescope of scarcely more than 8 inches in aperture; although continually seen in England and elsewhere through instruments of still less power; and although, by such aid, the surface of Mars has been mapped by harsh black lines in a manner that suggests the transformation of a world into a gigantic shunting-yard, the canals, at any rate in their generally assumed characteristics, have consistently refrained from appearing upon the picture of the planet formed in many of the finest telescopes in the world, directed by astronomers who, in other and independent work, have earned the highest reputation for keenness of vision. Through the Washington refractor of 26 inches of aperture, the instrument by which, in 1877, Professor Hall first detected the moons of Mars, the canals have never been traced. Dr. Keeler, of the Alleghany Observatory, made a special study of Mars when near its opposition in 1892 with a refractor of 13 inches of aperture. During the course of the observations the definition of the planetary outlines was frequently so excellent that the moons of Mars were clearly visible in the field of view; but although certain ill-defined shaded streaks were recognized near the recorded positions of canals, no trace of their hard rectilinear character, or of their marvellously reticulated system, was detected. Near the time of the opposition of 1894, Mr. Barnard, at the Lick Observatory, frequently directed the great telescope of 36 inches of aperture, the instrument by which he had already discovered the fifth moon of Jupiter, towards Mars. At times, when the seeing was most perfect, although the gray-green regions of the planet appeared richly covered by delicate and intricate detail, the very suspicion of which had never been suggested to other observers to whom the canals had been so startlingly conspicuous, features were, indeed, recognized upon the orange continents, but they were for the most part irregular, and consisted only of delicate gradations of light and shade. There was no appearance of hard, sharp lines. A few short, hazy streaks in the neighbourhood of the "Lake of the Sun" appeared as nearly the sole representatives of the Martian canals.

To explain the inconsistency apparent in these and other similar observations, it is not for a moment necessary to assume any want of good faith on the part of astronomers to whom the system of the Martian canals has appeared in all its wonderful complexity. Experience has fully shown, as every observer with the telescope has soon become painfully aware, to what a serious extent the eye may be deceived in its interpretation of details so faint as just to hover upon the verge of vision, and how readily unconscious bias, the result of even faintly preconceived ideas, may affect the judgment. Illustrations are abundantly supplied in the history of astronomical observation, and it will be sufficient to give three, selected almost entirely at random. In comparing recent photographs of the nebulæ surrounding the star η Argûs with the beautiful drawing of the same object made by Sir John Herschel during his residence at the Cape, differences of so startling a nature are found as to administer a severe shock to those who would put their trust in the eyes of man. The curiously definite border assigned, even by so careful an observer as Sir John Herschel, to a dark space in the nebula, known as the "key-hole", when compared with the perfect shading of light into darkness shown in the photograph, is an indication of the tendency of the eye to assign to excessively faint details a sharpness and a regularity that they do not possess. In inspecting sketches of the delicate detail of the Corona of the Sun, made at the same time and from the same place by different observers, it is frequently difficult to believe that the same object has been represented. Drawings of the Milky Way, as seen by the naked eye, have been recently executed by two independent observers, Dr. Boeddicker and M. Easton, each drawing the result of long and arduous observation, but, in comparing them, it is the exception rather than the rule to find any approximation in agreement in respect of the more delicate features.

Altogether it appears scarcely possible to avoid the conclusion that the existence of the canal system has not been established. There is no doubt, however, that in the course of a few years further light will be forthcoming upon the problem. At present the planet is becoming at each appearance less favourably situated for observation; but upon the return of favourable oppositions in 1907 and 1909, its disc will be scanned with an attention thoroughly aroused by the conflict of recent evidence, and with the aid of more powerful instrumental means than have hitherto been available. It might also be well if each observer should, before attacking the main problem, subject himself to a severe examination, in sketching through his telescope a number of illuminated distant discs, on which faint markings had been traced, but of a nature unknown to him. A personal tendency might be detected by the subsequent comparison of the drawings with the discs, which should serve as a valuable check upon his subsequent observations of Mars. By the comparison of a number of such carefully corrected records, it might be confidently anticipated that the riddle of the canals of Mars would receive its final solution.

It has frequently appeared a grave difficulty in interpreting Martian phenomena that the apparently mild climate, to which they have been generally thought to point as existing upon the planet, is inconsistent with its great distance from the Sun. There can be little doubt that if the Earth were removed to the distance of Mars, it would, by reason of the diminished intensity of solar radiation, become so much cooler, that nearly if not the whole of its oceans would be eternally bound in ice. Yet upon Mars the polar caps do not extend to lower latitudes than do those of the Earth, while the polar ice on the Earth is never reduced during the hottest summer to the insignificant remnant by which it is generally represented in summer upon Mars. From these facts it has frequently been assumed that the climate of Mars is actually warmer than that of the Earth.

Although it is not possible to make an exact estimate of the fall of temperature that would result if the Earth were removed to the distance of Mars from the Sun, a simple illustration will indicate its very serious extent. Taking the approximate ratio of 2 to 3 to indicate the relative distances of the Earth and Mars from the Sun, it follows that, since the intensity of radiation is inversely proportional to the square of the distance of the radiating body, the heating effect of the Sun's rays at every place upon the Earth's surface would be reduced by the square of $2/3$, that is, to $4/9$ of its present value. The heating effect of rays further depends, however, upon the angle at which they are incident upon the surface that absorbs them. The greater the obliquity, the less the heat developed upon equal areas, since the more slanting the incidence the larger the area over which the rays of a given columnar bundle would be distributed. In passing from the equator to either pole, for instance, the heating effect of the Sun's rays continually decreases, as the surface covered by bundles of rays of equal section increases. Let us now suppose the Earth to be at an equinox, and that it is regarded by an observer stationed upon the Sun. Imagine two parallel zones, each a mile in width, to be described entirely round the Earth upon its surface, one at the equator, and the other in latitude 63°. It can be shown by an exercise in elementary geometry that the second is, by reason of its lesser circumference, $4/9$ of the first in area, and that, as seen from the Sun, it appears from this cause, and also from its obliquity to the direction of vision, to be $(4/9)^2$ or $16/81$ as large as the equatorial one. This fraction then is the proportion between the angles subtended at the Sun by the zones, and it therefore also represents that of the quantities of heat received by them in equal times. The smaller zone therefore receives $(4/9)^2$ of the heat of the larger, but, as its surface is only $4/9$ as great, the heat received by a given area of the smaller zone is $4/9$ of that received by an equal area of the larger one.

But it has been shown that, if the distance of the Earth from the Sun were increased to that of Mars, the heating effect of the solar rays everywhere upon its surface would be reduced by this amount. Hence the equatorial heating effect upon the Earth, if it were transferred to the position of Mars, would be that at present found in a zone of 63° latitude. The parallel of 63° north latitude skirts the south of Iceland, it passes through Finland, and it traverses Northern Siberia, Alaska, the Hudson Bay Territory, and Greenland; and there is little doubt that the temperatures of these regions are upon the whole higher than would result from the direct radiation that they receive, since the Arctic and adjacent regions are warmed by currents of air from lower latitudes. Were the Earth, therefore, transferred to the distance of Mars, it might be confidently anticipated that, even at its equator, its climate would be of Arctic character.

The phenomena visible upon Mars do, however, suggest, though they do not probably demonstrate as conclusively as has frequently been assumed, that the temperature of the planet is very different from this, and attempts have been made to imagine some means by which the climate of the planet may be rendered less rigorous than its small allowance of solar heat would suggest. The possibility, one indeed that has never been seriously maintained, that the interior of Mars may be hotter than the Earth, and that its surface may be appreciably warmed by the outward flow of heat from its interior, may be briefly dismissed. It is probable that the interior of Mars, like the interior of the Earth, is hotter than the surface; but from the probability indicated by the nebular hypothesis, that Mars was developed in a highly heated condition before rather than after the Earth, and from the certainty that its cooling must, by reason of its smaller size, have proceeded far more rapidly, Mars should be the colder of the two planets. Since, in addition, measurements have shown that the heat conducted from the interior to the surface, even in the case of the Earth, is entirely insignificant in amount when compared with that received from the Sun, and is, therefore, a negligible quantity in directly affecting climate, it would appear impossible that the climate of Mars should be sensibly affected by the internal heat of the planet.

The only serious attempt that has been made to account for the assumed mild climate of Mars is based upon the property of selective absorption, exercised in some degree by all, and in a very marked manner by many, transparent substances. Selective absorption is illustrated to a remarkable degree in glass. If a plate of clear glass be held between the Sun and a thermometer, the bulb of which should be blackened upon the exterior to prevent the reflection of rays from the metallic surface of the enclosed mercury, it will be found that the indication of the thermometer is scarcely affected, the glass, transparent to light, being similarly transparent to the greater part of the rays that, upon their incidence on the blackened surface of the thermometer, develop heat. If, however, the same glass is interposed in the course of rays proceeding from a red-hot fire to the thermometer, a fall in temperature will be indicated, almost as great in amount as if an opaque screen had been substituted for the glass. Glass, therefore, is very transparent to the heat radiation of the Sun, but is practically opaque to that of a red-hot fire. It is to this last property that the efficiency of a glass fire-screen is due. Speaking generally, it is found that the higher the temperature of a body the more transparent is glass to the heating effect of its radiation; and this, not from the greater intensity of the radiation of a hotter body—for the experiment already described succeeds equally well even if, as may well be the case, the glass is placed so close to the fire that its radiation is, owing to its close proximity, more intense than that of the Sun on its arrival at the surface of the Earth—but by reason of some property impressed upon the radiation by the source. The wave theory of light, and of radiation in general, leaves little doubt that the difference in question is one of rapidity of the vibration of the ether of space as it transmits the waves that constitute radiation, but it is unnecessary here to go beyond the actual property of selective absorption as demonstrated by experiment.

The familiar fact that upon a clear day the air inside a greenhouse may be raised by the Sun's rays to a temperature far in excess of that outside, is frequently advanced as an illustration of the effects of selective absorption, though probably in this case the prevention of circulation of the enclosed air is partly responsible for the rise in temperature. The rays of solar radiation traversing the glass with readiness arrive at and are absorbed by the surfaces of the plants and other objects exposed to them. These, becoming heated in consequence, radiate their acquired heat in rays of essentially different character, which, being effectually absorbed by the glass covering, cause it to be heated by them. Consequently the interior, heated now both by radiation from the Sun and from the warmed glass, acquires a temperature which is frequently far in excess of that which would result if the glass had allowed a free path to the radiation from the objects within.

Tyndall has shown that a closely similar selective absorption may be exercised by many gases and vapours. He was unable to detect the property with certainty in dry air, but the presence of a very small quantity of the vapour of water in the air produced it in a marked degree. From the results of experiments arranged with considerable care and skill, he arrived at the conclusion that the vapour of water present in the atmosphere exercises an important influence upon the meteorology of the Earth, permitting the transmission through the air of the solar rays, but largely arresting the heat upon its return, by absorbing the radiation from the warmed Earth. It appeared, indeed, that under conditions in which the atmosphere contains an average amount of water-vapour in England, as much as 10 per cent of the Earth's total radiation should be arrested within 10 feet of its surface.

Based upon the selective absorption of water-vapour, the interesting speculation has been advanced that a mild climate upon Mars may result from the distribution throughout its atmosphere of water-vapour, in quantity so abundant that, by the efficiency of its trapping effect upon the solar radiation, it should more than atone for the great distance of the planet from the Sun. There is, however, considerable difficulty in imagining such a state of saturation of the Martian atmosphere as probable, or even possible. Although the physical conditions at the surface of Mars do appear to be in some respects favourable to the formation of water-vapour in its atmosphere, in others they appear to be extremely unfavourable; and though it is not possible to estimate accurately the opposing conditions, there can be little doubt of the unfavourable ones being the far more effective of the two.

The conditions upon Mars favourable to the existence of the vapour of water in its atmosphere, consist in the low intensity of gravitation at the surface of the planet, and the probable tenuity of its atmosphere. The amount of water that is capable of existence in an atmosphere in the state of vapour is entirely independent of the density of the atmosphere; experiment showing that although evaporation takes place more rapidly into rare air than into dense, yet the amount of vapour ultimately formed is the same, and is still the same even if the space into which evaporation takes place was initially a vacuum. In every case evaporation proceeds until the vapour immediately in contact with the evaporating surface has acquired a definite density, a density increasing with, and solely depending upon, the temperature, after which evaporation ceases. In the case of a planet, evaporation of surface-water would therefore continue until an atmosphere of the vapour of water had been formed that should, independently of any other atmosphere present, and therefore solely by its own weight, produce such a density in the vapour at the surface as would prevent further evaporation. Since the intensity of gravitation at the surface of the Earth is two and a half times that at the surface of Mars, an extension of water vapour two and a half times that necessary to prevent evaporation at the surface of the Earth would be possible in the atmosphere of Mars.

In the preceding argument it was necessary to assume a very simple condition—that evaporation should steadily continue until the whole atmosphere had become saturated. It is scarcely necessary to add, that, in the atmosphere of the Earth, this is very far from being the case. Owing to alterations in temperature in extensive bodies of air—due to different meteorological changes—condensation is continually occurring, resulting in the formation of cloud and mist, and the atmosphere of the Earth as a whole is, at all times, very far from being saturated. No doubt such condensation would also occur on Mars, but it is probable, from the low intensity of gravitation, that the meteorological changes would be less violent, and that condensation would therefore be less copious. Also, under conditions otherwise similar, evaporation would take place more rapidly into the rarer atmosphere, and the loss of vapour due to condensation would be more quickly restored. Perhaps the most definite form in which it is possible to express the general conclusion is—that if the intensity of gravitation upon the surface of the earth were reduced, and if, at the same time, the density of its atmosphere were diminished, there is little doubt that the atmosphere as a whole would be more richly charged with the vapour of water than it actually is.

In some respects, therefore, the physical conditions existing on Mars appear to be favourable to the formation of the vapour of water in its atmosphere. On the other hand, other conditions appear to be so extremely unfavourable that it is difficult to believe that these can be of much avail. Evaporation is the direct result of the radiation of the Sun acting upon the surface of water. Not only is the intensity of the solar radiation upon Mars less than one-half of its amount on the surface of the Earth, but the water surface exposed to it appears to be woefully restricted. The greater part of the Earth's surface is occupied by water. A nearly continuous tropical belt of ocean is exposed day after day to the direct radiation of a vertical sun. Upon Mars, on the contrary, the tropics are almost completely occupied by the orange continents. Gray-green regions, the aqueous character of which is more than doubtful, extend over the temperate zones. It is only in the arctic regions—the arctic regions of a planet whose tropics receive heat from the Sun that compares unfavourably in its amount with that received by lands of ice and snow upon the Earth—that there is, in the polar snows, any indication of water in either the solid or the liquid state. The faith of a keen believer in the habitability of Mars may see under such conditions an atmosphere heavily laden with moisture, but to us it appears that poor success has accompanied the attempt to warm Mars by a cloak of vapour.

Further, it appears certain that the trapping effect of the vapour of water has been much overestimated. If we accept Tyndall's estimate, that under average conditions in this country, 10 per cent of the heat radiated by the Earth is absorbed within 10 feet of its surface, it follows, as has been pointed out by Lord Kelvin, that so high a rate of absorption cannot continue; for if it did, 10 per cent of the heat escaping absorption in the first 10 feet being absorbed in the next 10, and so on, 90 per cent, or nearly the whole, would be absorbed in 200 feet, a conclusion that is directly contradicted by the very marked effect of cloud in checking the fall of temperature by radiation from the Earth's surface. It is probable that water vapour absorbs only a few waves of definite lengths among the many composing terrestrial radiation, that the very rapid absorption of these gives rise to the strongly-marked effect actually observed, but that the remaining waves, bereft of their more susceptible companions, escape without much further loss.

To reconcile the dissipation of the polar caps with the intense cold that it appears necessary to regard as prevailing over the Martian world, it has been suggested by Mr. Cowper Ranyard and other astronomers that the Martian snows may be the solidified form of some liquid other than water, and freezing at a lower temperature. The occurrence of carbonic acid gas as a constituent, howbeit a minor one, of the Earth's atmosphere, and the fact that by extreme cold it becomes condensed as a white powder, very closely resembling snow in appearance and melting at a temperature of about 120 Fahrenheit degrees below the freezing-point of water, has pointed to it as the origin of the polar caps on Mars. There are, however, very serious objections to this view. Under ordinary conditions of pressure, carbonic acid is incapable of assuming the liquid state, the solid upon being heated passing directly into the condition of gas. Under considerable pressure, however, the heated solid does melt, the resulting liquid boiling at a still higher temperature and becoming a gas. The least pressure necessary for this purpose is about five times that of the atmosphere at the surface of the Earth. Hence for liquid carbonic acid to exist on Mars, in consequence of the low intensity of gravitation, $12\frac{1}{2}$ times the mass of air must be accumulated over each square mile as is accumulated over a square mile of the Earth, an estimate that cannot possibly be accepted. If, therefore, the polar snows consist of solid carbonic acid, they must be formed by a direct precipitation of the hoar-frost of carbonic acid, and their disappearance must be a similarly direct process of evaporation. This conclusion is directly challenged by the appearance to Mr. Lowell and Professor Pickering of the blue-black belt fringing the disappearing cap of 1894, and the evidence that it furnished as to its liquid nature.

The actual deposition and dissipation of the hoar-frost of water is not inconsistent with a temperature considerably below the freezing-point, since direct evaporation takes place from ice at such low temperatures. Were it not for the evidence of the fringing belt, the gray-green regions might well be ice-bound seas, from which evaporation would take place under the cloudless Martian skies. The air would thus become charged to a slight extent with the vapour of water, which, distributed over the planet by atmospheric circulation, would be ultimately deposited as frost on the coldest polar regions.

In common with the greater number of other celestial objects to which it has become possible to apply a detailed examination, Mars has passed through a first stage in which it appeared a simple and an easy thing to interpret the features revealed, and has reached another, in which the first pleasing views have been rudely shaken, as observation has revealed difficulties at a far greater rate than it has solved them. Could we but traverse the millions of miles of planetary space that separate us from our ruddy neighbour, and dwell for a time upon its surface, we should look around us in vain for evidence of that fair miniature of the world we had left that formed the romantic picture of our fathers. It is more likely that we should find in Mars a succession of bleak arid deserts over which the rays of the vertical Sun would seem to struggle in vain to mitigate the blasting chill of the attenuated air. We should find, in higher latitudes, a succession of plains, clothed, perhaps, with elementary forms of vegetation capable of withstanding the rigours of a climate more than arctic in character. We should possibly encounter animal life, but assuredly in no familiar form. With the whole aspect of nature it would be difficult to associate romance, and we should be well content for the future to limit our acquaintance with the planet to the softened picture presented in the field of view of a telescope mounted on the more genial Earth.

Chapter IV.

The Analysis of Sunlight.

In the year 1672 Sir Isaac Newton published, among other discoveries in optics, the account of an experiment, in principle closely agreeing with one less perfectly arranged and interpreted by John Kepler more than half a century before, that was to form the foundation of a new branch of physics; one that, in its application to Astronomy a century and a half later, was destined to renew the youth of the oldest of the sciences, not unfrequently regarded then as approaching the termination of its active career and as having achieved its last triumphs. Newton's experiment was that of the analysis of sunlight, and the science that owes its origin to it is Spectrum Analysis.

In Newton's experiment, a circular hole was bored in the shutter of an otherwise darkened room, and through it, when the Sun was unclouded in the sky beyond, a beam of sunlight penetrated the room, and, following a straight course, formed an oval spot of white light upon the opposite wall. A prism—a block of triangular section—of glass was then placed in the path of the beam and immediately against the shutter, and, as the result, the white spot disappeared, and was replaced by a luminous band of coloured light considerably displaced from it in position. The band displayed from one end to the other a series of colours which closely corresponded with those seen in the rainbow, red appearing nearest the original position of the white spot, and violet at the end farthest from it.

From the fact of the displacement of the luminous image on the wall, the prism clearly exercised a deflecting effect upon the beam of light; and Newton accounted for the appearance of colour by the supposition that the light of the Sun was compound in nature, being, in fact, a mixture of all colours of the rainbow; that the combined effect of the whole upon the eye was to develop the sensation of white; but that the differently-coloured rays were deflected in different degrees in traversing the prism; red experiencing the least, violet the greatest, deflection, while the colours appearing between these were deflected to intermediate and different extents. The complete series he described as red, orange, yellow, green, blue, indigo, and violet; but it is probable that to most eyes the blue and indigo would appear as only different shades of the same colour. Newton supported his explanation by a number of simple though ingeniously-arranged experiments, and no doubt has since existed as to its soundness.

Fig. 7 represents, in simple diagrammatic form,

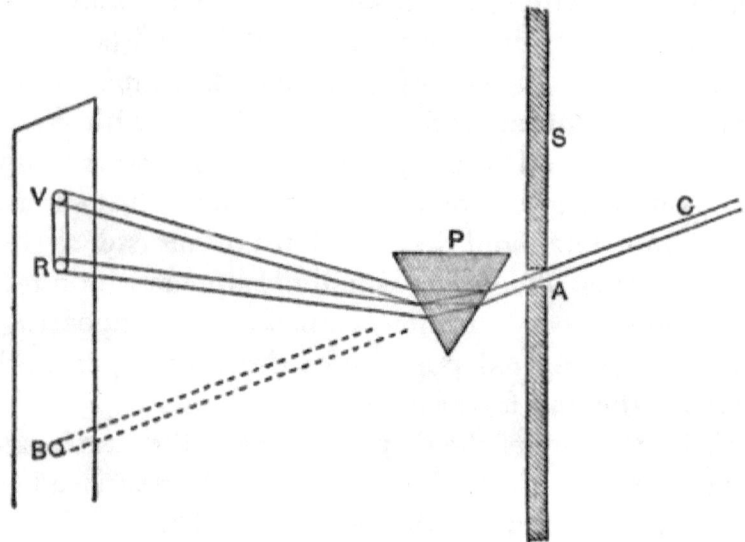

Fig. 7.—Newton's Experiment.

a vertical section of the arrangement of Newton's experiment, and requires but little explanation. S is the section of the shutter, A that of the circular hole. The Sun being in the direction of AC, a beam—that is, a bundle of its rays—arrives from along CA, enters the hole, and in the absence of the prism would traverse the room and form a white spot upon the opposite wall at B. Upon interposing the prism, however, at P, the rays suffer deflection, and are thrown upward, forming a coloured band between the limits R and V, the least deviated —the red rays—arriving at R, and the most deviated —the violet—at V, while the original white spot disappears.

The reader, if generally unacquainted with the first principles of optics, will probably find it well to trace the action of the prism on light in the more thorough manner developed in the next few paragraphs, but, should the above explanation appear entirely satisfactory and complete, these may be passed over.

It is a matter of common experience that light behaves normally—at any rate, to a very close degree of approximation—as if it travelled along straight lines. Were it not so, a distant object would not become hidden by the interposition of an opaque screen in the straight line between it and the eye. As a matter of fact, refined observations of the phenomena included under the term "diffraction" indicate that the transmission of light is not completely expressed in so simple a statement, but no error will arise from its assumption in the present instance. To light thus regarded as proceeding along a straight line, the term "ray" is applied.

A ray of light continues its path in a straight line, however, only for so long as the substance—or medium—through which it is transmitted is of absolute uniformity. It is a matter of common knowledge that, in passing from one medium into another, the course of light is deflected or "refracted" at the separating surface. Thus, in fig. 8, let PQ indicate the course of a ray in air incident upon the surface of water at Q. The path of the ray within the water will still be along a straight line QR, but this will not be the continuation of its former direction. It has been found that, in travelling from a rarer into a denser medium, a ray of light is commonly refracted towards the normal—or perpendicular — to the separating surface at the point of incidence; and that, conversely, in its passage from a denser into a rarer one, it is deflected from the normal. Thus, the ray PQ, if continued in its original direction, would proceed along QS; but, on entering the water, it is actually deflected towards the normal NN' into the direction QR. In travelling the reverse way, a ray, following the course RQ while within water, would, upon entering air, be deflected from the normal into the direction QP. Only a ray incident normally upon a separating surface penetrates it without experiencing refraction. From the refraction of light follows the familiar fact that an object in one medium, when viewed from another, generally appears to be in a direction different from that in which it really is.

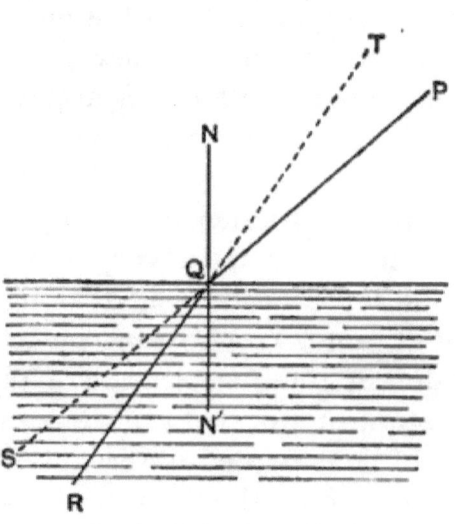

Fig. 8.—Refraction of Light.

The Analysis of Sunlight.

Thus, once more referring to the figure, one of the rays proceeding from an object situated at R would follow the paths RQ and QP in succession, and, should it enter the eye at P, the object will appear as if it were in the direction PS, since it is from this direction that the ray arrives. By analogous reasoning, an object at P, as seen by an eye situated below the surface of water at R, would seem to be in the direction RT.

The amount of deflection experienced by a ray in the act of refraction depends upon the angle at which it is incident upon the refracting surface, as well as upon the natures of the two media. The exact law by which it is determined was discovered by Snell about the year 1621, but its statement is unnecessary for the purpose of the present study. The fact of refraction, as well as Snell's law, are simply explained by the wave theory of light.

In fig. 9 the paths are traced of three rays traversing a transparent prism of glass, the amount of refraction being in each case calculated from Snell's law. The manner in which the rays are always deflected towards the normal upon entering the glass, and from the nor-

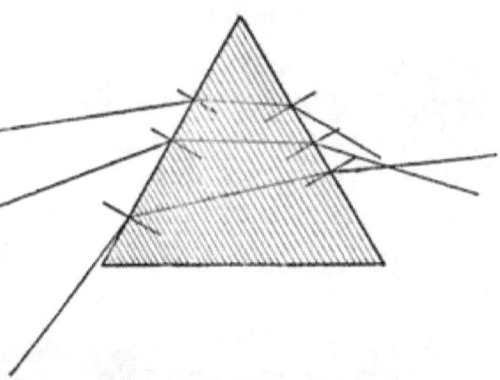

Fig. 9.—Paths of Rays traced through a Glass Prism.

mal on leaving it, should be carefully followed, and the construction of similar diagrams for other prisms of different vertical angles would form an instructive exercise. It is found that in every case the ray, by its passage through a prism, is deflected from the refracting edge, or that at which the two faces concerned in the refraction meet.

Experimenting with lights of different colours, it is found that they experience refraction in different degrees, red experiencing the least and violet the greatest deviation. In this lies the complete explanation of Newton's experiment. A very great number of colours and shades of colour are present in the light of the Sun. All of them coexist throughout the whole of the beam that enters the room. On traversing the prism, the violet rays in the beam are deflected far from the refracting edge, and by themselves would form a violet spot on the wall, largely displaced from the position of the original white one. The red rays would by themselves form a red spot, displaced to a less extent, while each colour and shade of colour would form a corresponding spot of coloured light between these two extremes. The coloured band—or spectrum—is therefore formed by a number of coloured spots, one formed by each colour and shade of colour present in the original light.

This explanation of the formation of the spectrum, although beyond doubt the correct one, is not free from difficulty, for it is truly wonderful that the very definite and distinct sensations of colour that are produced by the rays separately should so entirely disappear in the white that results from the excitement of all of them simultaneously. One of the most direct evidences of the soundness of the theory is that when, as may be effected by several simple methods, the colours of the spectrum are recombined, a perfect white, indistinguishable from the original, appears as the result of the mixture.

The number of colours represented in the spectrum of sunlight, as apparent to a normal eye, is generally regarded as seven, though it is probable that most observers would suggest six. Since the different colours owe their appearance to their being refrangible in different degrees, the experiment may at first suggest the view that there are in sunlight seven, and only seven, different kinds of light, each of a definite refrangibility. If this were the case, the spectrum would be formed of seven coloured patches. If the hole in the shutter were large, the section of the beam, and consequently each coloured patch, would be correspondingly large: and adjacent, or even non-adjacent patches, might overlap; but, by making the hole small, and so restricting the section of the beam, it should be possible to get rid of this confusion, since the angular separation effected by the prism would remain unchanged, while each coloured patch would decrease in size. Thus, let the row of seven black dots repeated in the first three lines of fig. 10 represent by their positions the amount of separation of the supposed seven different kinds of light in sunlight. Then, with a very small hole in the shutter, seven small, separate, and differently-coloured spots would appear, as indicated in the first line; while, by increasing the aperture, the size of the spots would increase and overlapping would occur, as represented in the second and third lines, a spectrum being formed, in which, as in the actual spectrum of sunlight, each colour would pass into the next by insensible gradations.

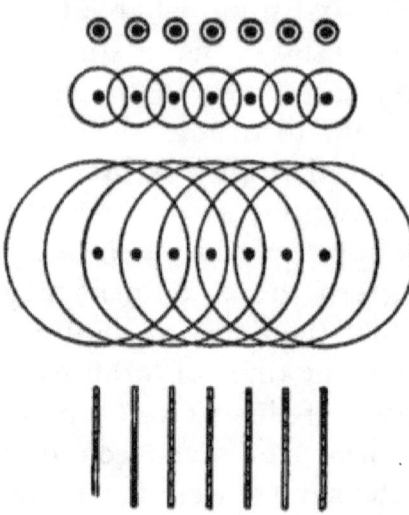

Fig. 10.—Formation of Pure and Impure Spectra.

That the perfect gradation of tint in the spectrum of sunlight is not due to the overlapping of only seven differently-coloured patches, is, however, shown by the fact that it is quite impossible to reduce the aperture to such dimensions that the spectrum shall be resolved into separate patches of colour. However small the aperture, and however distant the screen, the removal of which to a greater distance would have the same effect in tending to separate the coloured patches as reducing the aperture, the different patches pass the one into the other by perfectly insensible gradations. It is true that, with sunlight, breaks in the band might ultimately appear, due, as will be seen later, to another cause; but by substituting the light of a candle- or lamp-flame for that of the Sun, the spectrum would continue unbroken from end to end.

The impossibility of separating the coloured patches indicates that there are, for all practical purposes, an infinite number of colours in the light of the Sun or that of a candle. The description of the spectrum as consisting of seven colours indicates that in the gradual transition through the infinite series of rays from one extreme of refrangibility to the other, seven fundamentally different sensations are successively excited. The seven colours of the spectrum have reference to a physiological, not to a physical fact.

Instead of allowing the rays, after their separation by the prism, to illuminate a wall or other screen before being appreciated by the eye, they may be received by the eye directly; and, with most sources of light, this is the only means by which a sufficiently brilliant result can be obtained. The optical principles involved in this method of viewing the spectrum will be clear from the construction of fig. 11. Here A represents a small hole in a screen, supposed to be placed in front of a candle-flame, and E the eye of an observer, which should, however, be placed in practice immediately behind the prism. A ray of red light traversing a definite point in the hole will follow some such course as ABE, and will enter the eye. As the result, the eye will picture the position of A at some point, such as R, in the direction from which the ray arrived. The red rays traversing all points of the hole will behave in a similar manner, and the whole collection will appear as originating from a red circle at R. Such a reproduction of the appearance of the hole, a ghost from which the rays appear to come, is technically called its *image*. A violet ray traversing the same point in the hole, and accom-

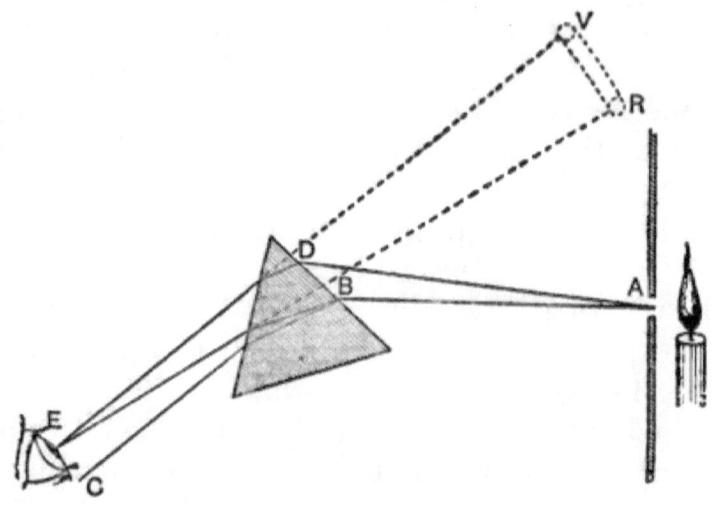

Fig. 11.—The Principle of the Spectroscope.

panying the original red ray along AB, will experience, in traversing the prism, a greater deflection than the red ray, and will be thrown into the direction ABC. It will, therefore, miss the eye altogether, and will be ineffective; but some other violet ray, such as AD, starting in a direction still more removed than the red ray from the direction of the eye, will, by its greater refraction by the prism, enter the eye, and will appear as if it came from V. The red rays, therefore, will develop in the eye an appearance of a red image of the hole at R, and the violet rays will similarly develop the appearance of a violet image of the hole at v. Every other colour present in the light will similarly develop a corresponding image of the hole in its own colour, in position intermediate between R and v, and the entire series, which will of course overlap, as when projected upon a screen, will form the spectrum of the light under examination. If the screen in front of the candle were removed, the red rays emitted by every point of the flame would give rise to the appearance of a red flame at R; the violet rays to that of a violet flame at v; and the other colours behaving similarly, there would result the appearance of a very impure spectrum formed by a great number of overlapping pictures of the flame in all the colours present in the light.

In 1802, Wollaston effected a great improvement in the method of obtaining the spectrum, by transmitting the light to be examined through a fine slit instead of through a round hole. The slit was arranged parallel to the edge of the prism. By this arrangement, the overlapping of images was very much reduced, or, in technical language, the purity of the spectrum was very much increased. Every colour present in the light would now be represented in the spectrum by a narrow band, which would become a fine line if the slit were sufficiently narrow; and these would overlap to a far less extent than the images of a round hole that should allow of the passage of an equal amount of light. If, for instance, under the conditions assumed in fig. 10, a slit were substituted for the hole, the appearance would be as represented in the fourth line, and the confusion occurring in the two immediately above it would be entirely avoided. It is of the highest importance to continually bear in mind that the appearance of the spectrum, whether formed by projection upon a screen or viewed directly by the eye, is due to a series of pictures or images of the aperture by which the light is admitted; a separate image being formed by each colour, or, more definitely, by each kind of light, as determined by its degree of refrangibility, represented in the original light.

Admitting sunlight through such a fine slit, and viewing the slit through a glass prism, Wollaston perceived the spectrum to be crossed at right angles to its length by four diffused dark lines. The lines were not seen in the spectrum of a candle-flame or with other artificial sources of illumination. A first glimpse was thus obtained of a discovery that was in a few years to revolutionize the study of the Sun and stars.

In 1814 further refinements were introduced by the celebrated instrument-maker, Fraunhofer of Munich. Fraunhofer's remarkable skill as an optician enabled him to construct prisms of finer quality and with faces more truly plane than had been found possible before, both points of great importance in giving accurate definition to the images produced; while, instead of viewing the spectrum directly, it was examined through a telescope, which received the rays immediately after their passage through the prism. As the result, the dark lines faintly seen by Wollaston became far more distinct, and their number was increased to 576. Some of them were also recognized in the spectra of planets and of fixed stars.

A further refinement, and one by which the prismatic spectroscope practically acquired its present form, was effected by Simms, another famous instrument-maker, in 1839. The image formed by a prism, or by the refraction of light at a single plane surface, is not, excepting under special conditions, sharply defined. This is due to the fact that, even with light of a pure colour, and, therefore, of only one degree of refrangibility, rays originating—and therefore diverging—from a point, do not after refraction diverge from any one point, a defect that arises from the particular form of the law of refraction and that we shall make no attempt to explain here. Since, owing to its dimensions, the eye receives not one ray of light but a number of rays; and since the refracted rays do not diverge from a point, the image appears slightly out of focus and indistinct. Light is diffused beyond what should otherwise be the sharp limits of the image, and overlapping of neighbouring images in the spectrum is unduly pronounced. If, however, the object is so far distant from the prism that the rays falling upon the face of the prism are sensibly parallel, all are refracted to the same extent, the emergent rays are still parallel, and indistinctness of the image due to this cause does not occur. Fraunhofer, who was the first to appreciate the importance of this condition, had been careful to approximate as close as possible to it, by placing the prism at a considerable distance—in some cases as much as 24 feet—from the slit; but the method was inconvenient, and involved a waste of light; since the greater number of the rays diverging from the slit missed the prism altogether. The improvement effected by Simms consisted in introducing a condensing lens in the path of the rays between the slit and the prism, at a distance equal to its focal length from the slit. The lens, known as the "collimating lens", collected the conical bundle of rays diverging from any point of the slit, and condensed them into a sheaf of parallel rays before they fell on to the nearest face of the prism. With a collimating lens, the slit need not, therefore, be farther from the prism than the focal length of the lens, which may be only a few inches. By these means distortion of the image due to the divergence of the incident rays was entirely obviated.

Reference must be made to one other condition of spectral purity, familiar to Newton and to later workers upon the analysis of light. It can be shown to follow from the form of the law of refraction, that, other conditions being the same, the definition of the image is least imperfect when the path of the ray in the prism makes equal angles with the refracting faces, a case illustrated by the central of the three rays traced in fig. 9. In practical work with the spectroscope, the prism is always adjusted for this to be, as nearly as possible, the case. The adjustment is made by turning the prism to and fro until it is observed that the spectral image is displaced from the direction of the slit to the least possible extent; optical theory having shown that this condition of "minimum deviation" is coincident with the desired symmetrical passage of the ray through the prism.

There should now be little difficulty in following the theory of the modern prismatic spectroscope, a diagrammatic section of which is given in fig. 12.

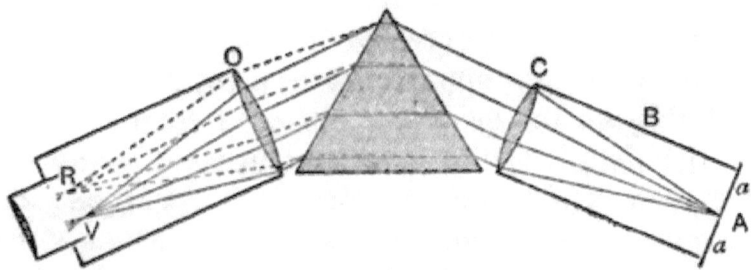

Fig. 12.—The Spectroscope.[1]

The slit is formed by bringing the carefully finished edges of two metal plates *a a*, almost into contact. By an adjusting screw, one of these can be moved towards or away from the other, by which means the slit, which is to be regarded as standing perpendicularly to the paper at A, can be made wide or narrow. The rays of light that enter the instrument by any point of the slit travel down a metal "collimating tube" B, usually from 1 to 3 feet in length, at the end of which they fall upon the collimating lens C, by which their paths are rendered parallel. Since everyone of them falls upon the first surface of the prism at the same angle, all

[1] The telescope and the collimator are usually from two to three times longer in comparison with their diameters than is represented in the figure, in which their sectional dimensions are exaggerated for the sake of clearness.

of the same colour are refracted equally, and their courses on traversing the prism, and after leaving it, are still parallel. Following now rays of one colour only, represented by the unbroken lines in the figure, they pass on to the telescope. The object-glass O, a condensing lens similar to that of the collimator, condenses then to a focus at V. On passing this they diverge again, traverse the eye-piece, and are received by the eye, to which, as the result, the appearance is presented of a line of light—an image of the slit—of a colour defined by the nature of the rays, and in position by the degree to which their courses have undergone deflection by the prism. Regarding, for instance, the rays, the paths of which have been traced, as violet; the red, represented by the dotted lines in the figure, experiencing less deflection, would be focussed in some such position as R, and other rays would be condensed in positions intermediate between R and V. As in the previous cases, a coloured line appears in the field of view for each kind of light radiated by the source. When the light has been sufficiently powerful to admit of a greater extension of its spectrum without undue enfeeblement of its colours, two or more prisms have sometimes been inserted between the collimator and telescope in such a manner that the light traverses all of them in succession. The separation effected is, of course, in direct proportion to the number of prisms employed.

Recently a "diffraction grating"—a surface of polished silver ruled very closely by a diamond with a number of parallel lines—has been frequently substituted for the prism. The light from the collimator falls upon the grating, and is thrown back, separated in the act of reflection into its constituent colours. A spectrum is thus formed, which is examined through a telescope, arranged nearly parallel to the collimator, and by the side of it. It is not possible to give here the explanation of the analysis of light by the diffraction grating, which is, however, in perfect accord with the wave theory of light. The separation of the colours is proportional to the closeness of the lines, and gratings recently constructed by Professor Rowland of Baltimore contain as many as 40,000 lines to the inch. The diffraction spectrum possesses certain advantages and disadvantages as compared with that formed by a prism.

In 1802 Wollaston detected four shaded lines intersecting the spectrum of sunlight at right angles to its length. Bearing in mind that the spectrum is, in reality, a series of pictures of the slit, one being formed by each kind of light present in the mixture submitted to analysis, it is clear that the shaded bands must indicate colours absent from, or feebly represented in, the light; as missing colours in sunlight. Three of the four lines Wollaston regarded as forming the natural divisions between different shades of colour—at best a not very satisfactory hypothesis, and one shortly to be disproved.

By the refinements introduced by Fraunhofer in 1814, the number of dark lines in the solar spectrum was increased to close upon six hundred, and it may be added that in a fine modern spectroscope upwards of ten thousand are visible. By rotating the observing telescope until the more conspicuous of them were brought in succession to the centre of the field of view, Fraunhofer was able, from the reading of a graduated circle attached to the telescope, to measure the positions of the lines in the spectrum, or, more definitely, the degree of refrangibility of each missing colour. From these measurements he constructed the first map of the solar spectrum, denoting the more conspicuous lines by the capital letters from A to H. The dark lines are still known as the Fraunhofer lines, and are identified by Fraunhofer's nomenclature.

The lines indicated by A and B upon Fraunhofer's map are situated in the deep-red of the spectrum, and are, under ordinary conditions, both strongly marked. C is a sharp well-defined line in the region of the spectrum where the red is passing into orange; D, a close pair of lines in the yellow; E, a condensed group in the bright-green; F, a sharply-defined line in the greenish-blue; G, a group in the deep-blue; and H, a rather wide pair of broad diffused lines, or rather bands, that lie in the extreme violet. This pair have assumed great importance in recent researches, and are now generally known as H and K.

For the Fraunhofer lines to appear in the spectrum, the slit must be very narrow. If it is opened beyond a certain degree of fineness, the images of the slit formed by the colours lying in the spectrum on either side of the position of the colour,

The Analysis of Sunlight.

to the absence of which the dark line is due, expand across its place and mask the effect of its absence.

Observations of the very impure spectra obtained by viewing flames directly through a prism had already been made by different observers with but little result, but now Fraunhofer subjected their light to examination in his more refined spectroscope. With no flames were any dark lines observed in the spectra, but bright lines frequently made their appearance, shining out conspicuously upon the background of a continuous spectrum. The bright lines clearly pointed to the presence, in the light under examination, of strongly developed pure colours that were not decomposed by the prism. In all flames that were examined, and especially strongly represented in the blue base of a candle-flame, were two nearly coincident shades of yellow, revealed by the appearance in the spectrum of two closely adjacent yellow lines. Fraunhofer remarked with astonishment that these yellow lines coincided exactly in their positions in the spectrum with the two components of the double dark line that he had already distinguished by the letter D in the spectrum of the Sun. The two shades of yellow, so abnormally abundant in the light of a candle, appeared, therefore, to be absent from, or, at any rate, but feebly represented in, the light of the Sun.

Later, in 1823, Fraunhofer examined the spectra of the brighter stars. In all of them he recognized dark lines, but although, in a few instances, the spectra appeared to be similar to that of the Sun in the distribution and in the relative intensity of the lines, in the greater number they were essentially different. The spectrum of Sirius, in particular, displayed only three dark lines, but all were far broader and more strongly marked than any recognized in the spectrum of the Sun.

For these as well as other reasons, Fraunhofer strongly maintained that the dark lines denoted colours initially absent in the radiation of the Sun and stars themselves, and that they did not originate, as had been suggested, either in the atmosphere of the Earth, or by some optical effect in the spectroscope, similar to that by which dark diffraction lines appear when an illuminated slit is viewed from some distance through a second slit, a phenomenon that was at that time attracting considerable attention.

The origin of the bright lines in the spectra of flames: the source of the dark lines in the spectra of the Sun and stars: and the remarkable coincidence established by Fraunhofer between the two yellow rays emitted by flames and the D pair absent in the light of the Sun, aroused the highest interest and evoked the keenest inquiry. For many years the exact study of the bright lines in flame spectra made but little progress. It was found that, upon saturating the wick of a flame with different chemicals, with the exception of the yellow pair that was always present, different sets of bright lines appeared in the spectrum. It appeared probable that the colours represented by the bright lines were emitted by the glowing vapours of the substances introduced into a flame, and that the continuous spectrum upon which they appeared was due to the normal radiations of the flame itself, for, with the scarcely luminous flame of burning alcohol, the continuous spectrum nearly disappeared, and the bright lines that flashed out upon the introduction of various chemicals seemed as if separated by intervals of almost complete darkness. The demonstration of the now familiar fact that each glowing vapour emits definite colours, indicated by bright lines occupying definite positions in the spectrum, and that from the appearance of its spectral lines the presence of an element may be inferred with certainty, was, however, only established by the classical researches of Bunsen and Kirchhoff in 1859.

But, in the meantime, facts of great interest were brought to light in connection with the dark Fraunhofer lines. In 1832 Sir David Brewster noticed that as the Sun approached the horizon many of the Fraunhofer lines became intensified. Several groups of lines towards the red end of the spectrum, for instance, while delicately defined when the Sun is high, appear at sunrise and sunset as massive black columns standing in front of the deep-red of the spectrum. Since, when at a low altitude, the rays of the Sun penetrate the atmosphere obliquely, and their path included in the air is therefore very great, Brewster suggested that those lines that were affected in this manner were caused by absorption by the Earth's atmosphere of the colours corresponding to them; their intensification with the low sun being due to increased absorption by reason of the greater atmospheric path of the light. The truth of this view has since been abundantly confirmed, and the lines that thus originate are known as *telluric* lines. The great majority of the Fraunhofer lines appeared, however, to be independent of the atmospheric track of the solar rays, since they were not affected in intensity by the altitude of the Sun, and it was therefore assumed that they denoted colours absent in sunlight before it entered the Earth's atmosphere. They were, in consequence, regarded as owing their origin to a similar absorption of definite colours in an atmosphere that was supposed to envelop the incandescent surface of the Sun.

In the discovery of the origin of the telluric lines, the first glimpse was obtained of the remarkable power possessed by many gases of absorbing colours so definitely as to more or less completely extinguish them without affecting those immediately on either side of them in the spectrum. In the following year Brewster showed that, instead of invariably necessitating an extensive atmosphere to produce the effect, with some gases a few inches were sufficient; for, on causing the light from a candle-flame to traverse such small lengths of certain gases before entering the slit of the spectroscope, the spectrum became ruled throughout by dark lines and shaded bands. Upon introducing a glass tube filled with the ruddy vapour of nitric peroxide between a lamp-flame and a spectroscope, the spectrum instantly became crossed by an enormous number of dark lines, some broad and massive, and others most delicately fine. Each dark line denoted the absence of a definite colour that had been absorbed by the vapour. Some of the lines appeared to coincide in their positions in the spectrum with certain of the Fraunhofer lines in the spectrum of the Sun, from which Brewster was led to conclude that nitric peroxide was a constituent of the Sun's atmosphere. Although this conclusion has been disproved, in suggesting it Brewster obtained the first glimpse of one of the most powerful and remarkable of the methods of modern scientific analysis.

The years that immediately followed Brewster's observations marked the birth of the science of photography. In 1838 Daguerre had discovered the process, with which his name has been since associated, for causing objects, by means of their light radiations, to impress their pictures upon specially prepared silver surfaces; in 1840 Dr. Draper had effected the first application of the discovery to astronomy in photographing the Moon; and two years later Becquerel succeeded in obtaining a photograph of the solar spectrum by projecting it upon a sensitive plate. In this first photograph of the spectrum the dark lines appeared as surely as in eye observation, while the remarkable fact became apparent, that the spectrum did not terminate with the violet, but extended beyond it to a distance far exceeding its visible limits, continuing, in its invisible extension, to be crossed by lines of absent radiation. It appeared, therefore, that the total radiation of the Sun contains rays more refrangible than violet light, and which do not possess the power of exciting the sense of vision. A year later, Draper, also by the aid of photography, similarly traced the solar spectrum beyond its visible limit in the red, and there also found Fraunhofer lines of absent radiation.

The time was now approaching when a successful attack was to be made upon the great mystery of the Fraunhofer lines. In 1849 M. Leon Foucault devised an experiment with the view of determining whether or not the coincidence—as regards position in the spectrum—between the two components of the Fraunhofer D line and the remarkable close pair of yellow lines that appeared in the spectrum of almost every flame, was exact. In this experiment, which has become classical, the yellow pair were obtained from the light of the electric arc. The electric arc is formed by passing a current of electricity across the space separating the ends of two carbon rods that are almost in contact. In passing through the rods the current experiences but little resistance, and therefore develops but little heat; but in its passage from one rod to the other across the air-gap separating them enormous heat is developed owing to the greatly increased resistance, and the air is raised to a very high temperature. The intensely hot bridge of air between the ends of the rods is technically known as the "arc". The temperature of the arc is so high that impurities present in the carbon rods, and indeed the carbon itself, volatilize and mix, in the state of gas, with the air in the gap. In spite of its high temperature, the arc itself gives but little light, owing to the poor radiating power of the gases forming it; but the ends of the rods, bathed in these highly-heated gases, are raised to the vivid state of incandescence that is the source of light in the arc lamp. The electric arc had been discovered by Sir Humphry Davy in the year 1800.

On directing the spectroscope toward either of the incandescent carbon ends, Foucault observed that the light agreed with that radiated by all incandescent solid bodies in yielding a continuous spectrum, but on deflecting the spectroscope towards the gap, so that the light from the glowing gases should be subjected to analysis, a number of separated bright lines were seen, indicating the existence, in the radiations of the glowing gases of the arc, of a corresponding number of isolated colours. Among the lines so seen the familiar yellow pair shone out conspicuously. To test whether these coincided exactly in their spectral position with the components of the D line in the solar spectrum, Foucault condensed the rays from the Sun upon the arc by means of an ordinary condensing lens. The solar rays, after traversing the arc, streamed onward, and entered the slit of the spectroscope along with the rays of the arc itself, and Foucault, anticipating that the coincidence between the positions of the yellow lines and the D lines would prove to be exact, confidently expected that, in the light of the Sun, poor in definite yellow rays, supplemented by that of the arc, abnormally rich in them, the D lines would altogether disappear.

On observing the spectrum, however, Foucault witnessed a most remarkable and unexpected appearance. Not only were the dark D lines not filled in by the yellow lines of arc spectrum, but they appeared to be both darker and wider than when the arc was absent. Not only did the radiations of the arc fail to supplement the deficiency of the similar radiations in sunlight, but the deficiency at once became more pronounced than before. Only one explanation appeared possible. Not only were the gases of the arc capable of radiating two definite shades of yellow, but they also possessed the power of absorbing them. The gases of the arc had absorbed a greater quantity of the yellow rays from the solar radiation than they had added to it, and increased darkness had been the result. Further, excluding the Sun altogether, Foucault, by the aid of a mirror, reflected the light from one of the incandescent carbon points through the gases occupying the gap between the pair; and on submitting the light thus transmitted to analysis, observed a continuous spectrum crossed by a pair of fine dark lines in the yellow. The arc had again absorbed the two shades of yellow more abundantly than it had radiated them, and the D lines had been produced for the first time in a laboratory experiment.

There can be little doubt that if the origin of the yellow pair had been known, the problem of the Fraunhofer lines would have found its solution in Foucault's experiments. The yellow lines had, however, proved a veritable stumbling-block to the advance of spectrum analysis. In the greater number of cases it seemed probable that the appearance of definite bright lines in spectra depended upon the presence of definite glowing vapours in the source of light, but the yellow pair seemed to defy any such limitation. They flashed out in the spectra of all flames, they seemed to be associated with the burning of all substances; and it was indeed suggested that they were developed in, and inseparably connected with, the process of combustion. For a few years after Foucault's observations they succeeded in evading the most refined methods of scientific inquiry. By the year 1852, however, Sir Gabriel Stokes had shown that they were absent from the spectrum of a candle-flame when the wick had been carefully snuffed clean and so as not to project into the luminous envelope, as well as from the spectrum of the flame of pure alcohol when burned in a carefully-cleaned watch-glass. On the other hand, they were most intensely developed when common salt—the chloride of sodium—and other compounds of sodium were introduced into flames. Gradually it became more and more probable that they were due to the glowing vapour of sodium, and that their almost universal appearance in spectra arose from the extreme difficulty of excluding a last trace of salt, and from their very powerful development upon the presence of the smallest possible quantity of it. Assuming this to be the explanation of their appearance, Sir Gabriel Stokes, in 1852, gave the correct explanation of the appearance of the D lines in the spectrum of the Sun.

Sir Gabriel Stokes's explanation was based upon theoretical grounds—the wave theory of light, and the view of the structure of matter involved in its acceptance. Since, in its later history, the most important applications of the analysis of light to astronomy have been directly due to the view of the nature of light indicated in the wave theory, it may be well to make a slight digression in a short sketch of its general features.

According to the wave theory of light—originally enunciated by Christian Huygens in the latter part of the seventeenth century, suppressed for a time by the overpowering authority of Sir Isaac Newton, but placed upon a sound scientific foundation early in the present century by the labours of Dr. Thomas Young—light is due to the transmission of waves, or undulations, from a luminous body to the eye. For there to be undulations there must be something to undulate, and to this something the name has been given of the "ether". To account for the phenomena of light, it is necessary to regard the ether, as not only existing throughout space, at any rate to the farthest of the visible stars, but as permeating all matter.

The display of iridescent colour frequently so exquisitely developed in light reflected from thin films, such as the envelope of a soap-bubble; the coloured fringes visible upon either side of an illuminated slit when viewed through a second and similar slit at a moderate distance from it; and the spectrum formed by a diffraction grating, enable us, if we interpret them according to the wave theory—by which alone they have so far received a satisfactory explanation—to measure the lengths of the ether waves. Estimates deduced from the different phenomena are in perfect accord, though there is no doubt that the highest degree of accuracy is obtainable in observations of the diffraction spectrum. From them it appears that colour, and therefore refrangibility, is determined by the length of waves in free ether; the sensation of red being excited by the longest and that of violet by the shortest waves that affect the eye, while the passage up the spectrum from red to violet is accompanied by a continual decrease of wave-length. The actual wave-lengths are about a sixty-thousandth of an inch for violet, and a thirty-thousandth of an inch for red light, but they are more accurately given in the table on p. 175.

It will scarcely be necessary to remind the reader that the appearance of the transmission of matter by wave motion is illusory. On the surface of water disturbed by wave motion floating objects merely rise and fall as the waves pass, which shows that the movement of the wave-conveying medium consists of a succession of oscillations up and down, while the waves themselves continually pass them horizontally. The illusion of water moving with waves results from each portion of the surface transmitting its disturbance to the portion immediately in front of it, but occupying a definite interval of time in so doing. After a short interval, therefore, the form of the water surface has been moved forward, but so continuously that the appearance is produced of the surface itself having been displaced in the direction of the wave motion.

The double appearance of objects when viewed through Iceland spar and other crystals, as well as the chromatic effects and general properties of polarized light, indicate that the motion of the ultimate parts of the wave-conveying ether is transverse, or across the direction of wave motion. In this respect ether waves resemble waves upon the surface of water, as well as those upon stretched strings. They differ in character from waves of sound, in that in these the motion of the air—the undulating medium in their case—consists of oscillations to and fro *in* the direction of wave motion.

During the passage of a single wave past a point in the ether, the ether at the point executes a single vibration or oscillation about its normal position, this vibration being, according to the inference of the preceding paragraph, across the direction of wave motion. During the passage of a train, or series, of similar waves, the ether, therefore, continues to oscillate, and the number of oscillations executed every second—a quantity known as the "frequency of oscillation"—is determined by, and is equal to, the number of waves passing in a second. Since all waves travel with the same speed, the longer will pass in less rapid succession than the shorter, and will therefore produce a less rapid oscillation. The violet waves, for instance, being only about half as long as the red, are associated

The Analysis of Sunlight.

with double the frequency of oscillation. From the geometry of wave motion thus sketched it follows from simple reasoning that in every case the velocity of the waves is equal to the product of their length into the frequency of oscillation. From this relation it is possible to determine the frequency of oscillation of different kinds of light waves, since their velocity—the speed of light—is known, and their length may be determined in every case by a diffraction grating. This has been carried out in the following table, the speed of light being taken as 187,000 miles per second. The way in which the frequency of oscillation increases as the wave-length decreases should be carefully noticed.

WAVE-LENGTHS AND FREQUENCIES OF OSCILLATION OF ETHER WAVES.

Colour.	Fraunhofer Line.	Wave-length in free ether,[1] in millionths of an inch.	Vibrations per second in millions of millions.
Red	A	29·9	395·8
	B	27·0	437
Orange	C	25·9	458
Yellow	D	23·2	510
Green	E	20·7	570
	F	19·1	618
Blue	G	17·3	683
Violet	H	15·6	757

[1] The necessity for the addition of the words "in free ether" is due to the fact that, while traversing transparent substances, the speed of light is reduced, doubtless as the result of the close association of the ether with ordinary matter. The frequency of oscillation must, however, remain the same, being always that of the source of the waves, so that the relation—speed of waves = wave-length × frequency—indicates that the wave-length must be reduced in proportion to the speed. It is customary to define a particular kind of light by its wave-length. Since, however, this varies with the medium through which the light travels, it would be far better to define

The most simple view to take with reference to the generation of waves in the ether, is that the luminous—or wave-generating—body contains oscillating portions of matter that possess a grip upon the ether. As the oscillations of a hand or tuning-fork may develop waves upon a stretched cord, so these vibrating parts, gripping the ether, and being thus able to transmit their movement to those portions of it immediately round them, set up waves, the frequency of oscillation of which is the same as their own. What these oscillating parts are is not of fundamental importance to the present study, but they are generally regarded either as the atoms of matter, or as portions of, or structures in, the atoms, elastically attached and capable of oscillation within them. The atom of sodium, capable of emitting two yellow rays of nearly the same tint, that is, of developing two series of waves of nearly the same frequency, may be regarded as analogous to a musical instrument with only two strings tuned nearly to unison. Since the waves generated by the incandescent vapour of sodium cause the ether to oscillate about 510 millions of millions of times in a second, this rate of vibration must also be that of the structures themselves. Below the temperature at which the yellow light is emitted we must suppose that the structures are not oscillating with

---

it by its frequency of oscillation. When light is defined by its wave-length, that in free ether should always be understood. In glass the speed of light is about ⅔ of its speed in free ether. In air it is scarcely affected. Refraction is the direct result of alteration of speed in passing from one medium into another, and from the extent of refraction the alteration in speed, and therefore in wave-length, can be determined.

this frequency, but that the oscillations may be developed by sufficient addition of heat. The higher the temperature, the more intense the radiations, and therefore the more intense the oscillations of the structures. The reader is probably aware that there is good reason for regarding such atomic and molecular vibration as constituting the sensible heat of matter.

Adopting this view as representing the mechanism of radiation, that of absorption follows naturally. Upon a series of ether waves traversing a space throughout which atoms of matter are distributed, the atomic structures gripping, and therefore gripped by, the ether, will tend to be thrown to and fro in harmony with its movement. Heat will be represented by the motion generated in the atoms, while the energy of the waves themselves will be correspondingly decreased by the loss of the motion transferred from them to the matter.

There is, however, one case in which the transference of the energy of motion from the ether waves to the atoms will be especially pronounced. It is a familiar fact, deducible from the first principles of mechanics, that, if a body capable of independent vibration is acted upon by a succession of impulses acting in unison with its own oscillations, a far more extensive oscillation will result than if such a coincidence did not exist. The principle, generally known as that of sympathetic vibration or resonance, is abundantly illustrated throughout mechanics and physics. If a weight be suspended by a cord, the upper end of which is held in the hand, a succession of properly timed, but scarcely appreciable, movements of the hand to and fro may cause a very extensive oscillation of the weight. If a large man be seated in a garden swing, a little man, by properly timing his thrusts into unison with the oscillations of the swing, may develop in the large man an oscillation out of all proportion to that which would otherwise result from his most violent efforts. A musical note sounded in the neighbourhood of a jar of such dimensions that the natural period of oscillation of the contained air coincides with that of the note, will cause the air of the jar to sound loudly in response. The recently discovered possibility of electric signalling over considerable distances without connecting wires depends upon the coincidence between a succession of feeble electric impulses applied to a distant conductor, and the normal oscillations of electricity in the conductor.

If, now, white light—that is, a number of wave-trains of all possible frequencies between the limits of the spectrum—should traverse the vapour of sodium, it should not be difficult to predict what would occur. Those waves, the frequencies of which did not agree with the natural vibrations of the sodium atoms, would scarcely affect them, and therefore they themselves would be scarcely affected. Those waves, however, that possessed vibration frequencies in unison with the normal oscillations of the atoms, would apply impulses to the atoms or atomic structures accurately timed to their own oscillations; resonance would follow, and extensive motion would be developed in the atoms. During the development of this motion, equal energy of motion would be absorbed from the waves. The waves, from which this energy would be absorbed, would be damped, and the light, after having traversed the vapour, would be found deficient in precisely those waves that the vapour itself could originate. Generally any vapour possessing the power of emitting definite radiations must also possess a special capacity for absorbing them.

The deficiency in sunlight of the two shades of yellow emitted by the glowing vapour of sodium, indicates, therefore, that the white light of the Sun has traversed the vapour of sodium somewhere in its passage to the surface of the Earth. Since the vapour of sodium is not found in the atmosphere of the Earth, and is assuredly not distributed in interplanetary space, it must be looked for in the atmosphere of the Sun.

Such was Sir Gabriel Stokes's explanation of the double line D, and of other Fraunhofer lines, though up to that time no other coincidence between Fraunhofer lines and bright lines in the spectra of glowing terrestrial vapours has been established. With the singular modesty and reticence that has characterized him through life, Stokes, while offering one of the most remarkable of scientific theories as a suggestion in private conversation with a friend, refrained from making it public. In the following year (1853), however, a similar explanation was given by Ångstrom of Upsala. Ångstrom, moreover, established the coincidence between other of the

Fraunhofer lines and certain bright lines, though of unknown origin, in the spectrum of the electric arc.

In 1859 the publication of the classical researches of Bunsen and Kirchhoff placed spectrum analysis upon a sound foundation as a branch of science. For the first time, bright lines in the spectra of flames were definitely proved to arise from the presence of glowing vapours in the flames. The flame generally employed was that of a spirit-lamp, or of gas which had been deprived of its luminosity in the now familiar form of Bunsen burner. A great number of different substances were made to pass into the flame in the state of vapour, by introducing them in the solid or liquid state upon a piece of platinum wire into the lower part of the flame. It was found that every system of bright lines was associated with the presence of a definite vapour in the flame, and with such consistency that the presence of the vapour could be inferred with certainty from the appearance of its characteristic lines in the spectrum. This fact, of course, constitutes the very foundation of spectrum analysis. The famous close yellow pair were traced to sodium; and it was shown that their continual appearance in all sorts and conditions of flames was due to the universal distribution of common salt in the atmosphere, carried into it in all probability in the first instance by sea spray, and to the marvellous delicacy of the spectral test, a delicacy so extreme that the yellow lines appeared in the spectrum on the introduction of a two-hundred-millionth part of a grain of salt into the flame.

Not only was the complete set of bright lines yielded by one glowing vapour different from that given by any other, but only rarely did any one of the lines of one element appear to occupy in the spectrum the position of a line of another. Whether the few coincidences that have to the present time been observed are in any case more than approximate, the result of insufficient spectroscopic power; whether they are exact; and, if so, whether they are more than accidental, has been very keenly discussed in recent years in connection with certain modern speculations. In a great many instances apparent coincidences have been shown, by the application of more refined instrumental means, to be only approximate, and the view is now generally held that the few as yet unresolved will either yield to higher dispersion or are merely accidental.

In 1859 the splendid results obtained by Bunsen and Kirchhoff were brought to bear upon the problem of the Fraunhofer lines by Kirchhoff. Kirchhoff found no difficulty in obtaining the sodium lines dark upon the background of a continuous spectrum, by interposing the flame of a spirit-lamp, upon the wick of which a few grains of salt had been sprinkled, in the path of rays proceeding from the incandescent lime of the limelight to the slit of a spectroscope. Further, he failed to obtain the effect when the flame of a bunsen burner similarly charged with salt was used instead of the spirit-lamp, but perceived instead the bright-yellow pair radiated by the flame superposed upon the less brilliant continuous spectrum of the lime-light. From this he suspected that to effect "reversal" the temperature of the vapour must be less than that of the radiating source.

The statement of the exact conditions under which a vapour will effect the reversal of its spectral lines was first given by Balfour Stewart in 1861. In 1791 Prevost of Geneva had published a suggestive paper entitled "On the Equilibrium of Heat", in which, according to a law then enunciated for the first time, and since known as "the law of exchanges", it was shown that every body when at the same temperature as those surrounding it, must possess a power of absorbing heat radiations in direct proportion to its power of emitting them. At the time of Kirchhoff's researches, it was thought to be probable, from the similarity in the laws by which they were governed, that radiant heat and light were different forms of one kind of radiation, and Balfour Stewart perceived that the extension of Prevost's reasoning to light radiation would account, not only for the fact of reversal, but also for the condition, already suggested by experiment, that to effect it, the absorbing vapour must be cooler than the source.

Limits of space, unfortunately, make it impossible to introduce Balfour Stewart's reasoning here, but an excellent outline is given in Balfour Stewart's *Heat*, and will well repay the most careful study. By a very simple process of reasoning, based upon the obviously sound assumption that a body within an opaque enclosure, all portions of the inner surface of which are at the same constant temperature, will ultimately acquire the temperature of the enclosure, Balfour Stewart showed that such a body, when at the temperature of the enclosure, must absorb and emit any particular kind of radiation in exactly equal amount. Since the rate of radiation from a surface is directly dependent upon temperature, while the rate of absorption depends upon the nature of the surface and is not directly affected by its temperature, it follows that, under the conditions imagined, any fall in the temperature of the body will cause its radiation to fall short of its absorption, while a rise in temperature will cause its radiation to exceed the absorption.

From this conclusion it is possible to state definitely an essential condition for the appearance of dark lines in the spectrum of the Sun. It is, that the gases present in the solar atmosphere must be at a lower temperature than the incandescent surface—or photosphere—behind. If the temperature of the gases were equal to that of the photosphere, they would be in such a condition as to absorb from its light precisely as much of any kind of radiation as they would add to it, and, in consequence, the light from the photosphere would, after traversing them, be unchanged in composition. If the temperature of the atmosphere were to fall below that of the photosphere, the radiation of its gases would, for each particular kind of ray, fall short of the absorption, and dark lines would result in the spectrum; while, if the temperature of the atmosphere were to rise above that of the photosphere, the radiation of its gases would exceed the absorption exercised by them, and bright lines would appear upon the continuous spectrum of the photosphere. From the general appearance of dark lines, the Sun's atmosphere may be assumed to be cooler than the photosphere, though, as will be seen later, bright lines do occasionally appear.

It is clear from these principles that the Fraunhofer lines are not absolutely dark, but only appear so by contrast with the more brilliant spectrum of the photosphere upon which they are projected. Even if a gaseous constituent of the solar atmosphere were a perfect absorber of particular kinds of light, its own radiation, which would be of the same nature as the absorbed light, would travel on with the photospheric rays that had escaped, and would in part supply the place of those that had been absorbed. There is no doubt, that if the incandescent surface of the Sun were for a moment to be extinguished, while its atmosphere remained unaffected, the solar spectrum would appear as a crowd of bright lines corresponding to actually existing dark ones. The existence of the photosphere behind does not detract from the light that we receive from the atmosphere, but, by filling the spaces between its bright lines with more intense light, causes them to appear dark by contrast. That the apparently dark Fraunhofer lines are in reality brilliant can be shown by carefully-arranged experiments.

The presence of sodium in the atmosphere of the Sun having been established, Kirchhoff next endeavoured to discover other of its constituents, by searching in the spectra of terrestrial elements for bright lines that should coincide in their positions with Fraunhofer lines. His efforts were entirely successful. On passing the intense electric discharge of an induction coil from one metal wire to another across a short air gap, a brilliant spark was obtained, which, when analysed, gave a spectrum of bright lines—clearly due to the glowing gases filling the gap. Different sets of bright lines appeared as different metals were employed to form the spark, and it was clear that the spectrum consisted of the bright lines given by incandescent air, together with those due to the glowing vapours of the metal wires, these being partially volatilized by the intense heat of the discharge. Passing the discharge between the ends of iron wires, the spectrum given was that of a mixture of the vapour of iron and air, and Kirchhoff succeeded in establishing coincidences between no fewer than sixty of the lines that were due to the iron and dark lines in the spectrum of the Sun. It was, therefore, proved that the vapour of iron was a constituent of the atmosphere of the Sun.

Continuing his researches by this method, Kirchhoff considered that he had demonstrated the existence in the atmosphere of the Sun of nine metals known to terrestrial chemistry. They were—sodium, iron, calcium, magnesium, nickel, barium, copper, zinc, and chromium. Further, it was regarded as demonstrated, from the absence of their characteristic lines, that twelve other metals, including gold, silver, and mercury, were absent.

In 1862 Ångstrom published the results of an extensive series of observations, similar in principle to those of Kirchhoff, on the chemistry of the Sun's atmosphere. Experimental details differed from those adopted by Kirchhoff in that the analysis of the light was effected by the diffraction grating, and in the substitution of the electric arc for the discharge of an induction-coil as the source of heat. The results were in general agreement with those of Kirchhoff, but a few additional elements were detected, among which the most interesting was hydrogen. The spectrum of hydrogen is of the highest importance in the astronomical applications of light analysis. When enclosed within a glass tube, under a pressure considerably less than that of the atmosphere, and subjected to a discharge of electricity—generally led into and from the gas by platinum wires penetrating the glass—hydrogen gas becomes luminous, emitting a soft peach-like glow. In 1859, Plucker, subjecting this glow to analysis in the spectroscope, had found that it consisted in the main of three bright colours, represented in the spectrum by three bright lines—the first of a magnificent crimson, the second of a bluish-green, and the third of a deep-blue colour. Ångstrom found that each one of these had its counterpart among the Fraunhofer lines, and in 1866 he detected a fourth line—of a violet colour—in the hydrogen spectrum, and found that it also was represented by a dark line in that of the Sun. The important lines c and f in Fraunhofer's nomenclature are the reversals of the crimson and bluish-green lines of hydrogen.

It is unnecessary to give more than the briefest

outline of the later history of these methods. In 1872 Sir Norman Lockyer commenced a laborious series of comparisons of photographs of the solar spectrum with those of spectra of metals volatilized and rendered incandescent by the electric discharge of an induction-coil, and, as the result, succeeded during the following four years in adding about twenty new elements to the fourteen that had been previously recognized in the atmosphere of the Sun. In 1887 Messrs. Trowbridge and Hutchins demonstrated the existence in the Sun of the vapour of carbon, the first element of a non-metallic nature that had been found in its atmosphere; while in 1891, also from photographic comparisons, silicon was detected by Professor Rowland. Some idea of the fulness of detail shown in Rowland's photographs may be gained from the fact that coincidences were established in them of upwards of two thousand of the Fraunhofer lines and bright lines in the spectrum of iron.

By the series of researches that have been traced, culminating in the work of Kirchhoff, spectrum analysis was raised into the position of an exact science. It appeared to be all-powerful in problems to which the application of its methods was possible. Every glowing gas was regarded as emitting and absorbing definite radiations. By the appearance of their radiations in the spectroscope gases could be detected with certainty; and it was at first not unnaturally concluded that by the absence of their radiations from glowing matter their own absence could be asserted with equal confidence.

Had this anticipation been realized, the determination of the presence or absence of any terrestrial element or compound in the atmospheres of the Sun and stars would have been only a matter of careful and sufficiently prolonged observations, and the later course of physical astronomy would have been strangely different from its actual history.

From about ten years after the date of Kirchhoff's work, it has become increasingly apparent, that, although in every case the presence of a glowing gas is directly demonstrated by the appearance of its characteristic radiations, some caution is necessary before the absence of the gas can be inferred with equal certainty from the absence of those of its radiations with which we are familiar. It has been found that change of physical condition, change that may result from alteration of temperature or pressure, or even from the admixture of other substances, may cause the familiar radiations of a gas to disappear and to be replaced by others, frequently to such an extent that its spectrum may assume an entirely new and unfamiliar character, in which no relation to its former self is apparent. Of the many instances of such modifications of spectra that have been studied, attention may be specially directed to two, both of great importance in astronomical physics.

We have seen that the visible spectrum of hydrogen consists in the main of three bright lines—a crimson, a green, and a blue—while there is a deeper violet fourth that appears under strong electrical excitement. Of these, the crimson is the one that appeals most strongly to the eye. In 1869 Sir

Edward Frankland and Sir Norman Lockyer made some remarkable observations upon these lines and those of nitrogen. Hydrogen gas was first enclosed in a glass tube through which an electric current was transmitted, and was then rarefied by the action of an air-pump connected with the tube. The gas within the tube became incandescent under the influence of the current, and as rarefaction proceeded, the lines of its spectrum became finer and more brilliant, and after a time a limit was reached at which the first three were most conspicuous. The gas was then subjected to a further rarefaction while the electric discharge was maintained at a moderate intensity. During the progress of the rarefaction the spectrum was carefully observed, and it was seen to undergo a striking alteration. The crimson and blue lines became fainter, although the green line was scarcely affected, while, ultimately, the crimson and blue entirely vanished, leaving the still strong green line as the sole representative of the radiations of hydrogen. The more complicated spectrum of nitrogen was, under similar conditions, also reduced to a single green line; and, as might perhaps have been expected, a mixture of hydrogen and nitrogen under these conditions yielded, when traversed by an electric discharge, a spectrum consisting of the two green lines already observed. But, by now moderating the discharge so that the temperature of the gases should be reduced, the green nitrogen line in its turn disappeared, while that due to hydrogen still shone out conspicuously; so that, in a mixture known to contain nitrogen, and traversed by a current of electricity sufficient in intensity to cause a gas mixed with the nitrogen to glow, no trace of nitrogen was recorded in the spectrum.

Changes in the spectrum of calcium are no less remarkable or important. Compounds of the metal calcium—the metallic base of lime—when introduced into the flame of a bunsen burner by the means already described, cause the flame to acquire a brick-red tinge. Observation of its spectrum shows this light to be composed of rays of different colours, conspicuous among which is red—represented in the spectrum by a broad red band. When introduced into the electric arc, the temperature of which is considerably higher than that of the bunsen flame, the vapour of calcium gives a spectrum in which the red band has become much reduced, while a strong blue line, invisible in the flame spectrum, has appeared, as well as a pair of brilliant violet lines. In the spark from the induction-coil, which is probably at a still higher temperature, the spectrum entirely loses its red ray, the blue becomes fainter, while the violet pair are far more strongly developed than before. The last spectrum of calcium is, therefore, as different from the first as if two different metals had been subjected to examination. Passing now to the Sun, we find among the Fraunhofer lines the reversed images of the lines of the last—the spark spectrum: H and K, the great dusky pair lying almost at the limit of the violet, corresponding in their positions with the two broadened violet lines, and a fine dark line to which no special name has been given with the blue line. It is further interesting to notice that in light condensed from solar prominences—which are generally regarded as hotter than the general atmosphere—upon the slit of the spectroscope (the method by which this is effected will be described in a later chapter), the blue line has in its turn disappeared, and the strong pair, H and K, alone remain as the representatives of calcium.

Sir Norman Lockyer has interpreted the change in the spectrum of calcium, as well as similar instances presented by spectra of other metals, as the direct effect of increase in temperature, and maintains that they lend strong support to the view, of which he has made himself the champion, that by increase of temperature terrestrial elements become "dissociated" or resolved into still more elementary forms of matter. According to this view the pair of violet lines are not radiated by the vapour of calcium, but by the vapour of some element contained in calcium and dissociated from it by the temperature of the electric arc and that of the atmosphere of the Sun, while the substance emitting the red rays displayed in the bunsen burner has been entirely decomposed at these temperatures. Similarly, the substance, the radiation of which contains the blue line, is first dissociated from calcium at the temperature of the arc, is partially dissipated in the hotter spark, and is entirely destroyed in the still more intense heat of the prominences. In 1897, however, Sir William Huggins succeeded in effecting the same changes in the spectrum of calcium by reducing the density of the vapour, without, as he confidently believed, the change being accompanied by any appreciable increase in temperature, so that it appears probable that in the experiments first described the changes appearing on increase of temperature may have been only indirectly due to that cause, the direct influence having been the reduction in density consequent upon the expansion of the heated vapours.

Although all of the more conspicuous of the Fraunhofer lines have now been connected with bright lines in the spectra of terrestrial elements, the great majority of the whole are still unidentified. It is, of course, possible that many or even all of these may yet be found to correspond with the bright lines of terrestrial elements if it should become possible to produce in the laboratory conditions more closely approximating to those that exist in the atmosphere of the Sun. The failure to detect the faintest trace of absorption by oxygen in the solar radiations is very remarkable, in connection with the extensive distribution and supreme importance of oxygen in the atmosphere and in the crust of the Earth. It is true that oxygen is magnificently represented among the Fraunhofer lines, the great groups A and B being due to it, but there is no doubt that these are entirely caused by absorption in the atmosphere of the Earth. They become decreasingly conspicuous in the solar spectrum as observations are made from higher and higher altitudes, and at such a rate as to indicate that beyond the farthest limits of the atmosphere all trace of them would disappear from the radiations of the Sun. It is of course still possible that oxygen may exist in the Sun's atmosphere under physical conditions so differing from those with which we are familiar that its spectrum is unrecognizable, or it is conceivable that it exists dissociated into other and more elementary forms of matter, the uninterpreted record of which is before us among the many thousands of the Fraunhofer lines whose language has still to be read.

Chapter V.

The Analysis of Starlight.

In the previous chapter we have traced the succession of sure though laborious steps, by which, from the decomposition of a beam of sunlight in Newton's study, a new science has been constructed, that has given us a revelation of the chemistry of a body nearly a hundred million miles away across apparently empty space, the very suggestion of which would have seemed utterly preposterous a hundred years ago. That a beam of sunlight, less than a thousandth of a square inch in section, should contain latent within it the record of the constitution of the Sun's atmosphere may well induce hesitation in imagining any limit to the powers of scientific methods. While, moreover, steadily extending astronomical discovery in this direction, the new method has been developed with no less astounding success along other lines. It was considered advisable to ignore these for the time, and they will form the subject of the present and the following chapter.

We have seen that Fraunhofer in 1814 directed to the light of the stars the method that he had already applied with such remarkable results to that of the Sun; and that he found their spectra to be crossed, like the spectrum of the Sun, by a number of dark lines. He found, moreover, that while in some cases the dark lines of stellar spectra agreed closely with those seen in the spectrum of the Sun, they were more often different from them both in their positions and their relative intensity. The method adopted by Fraunhofer in these investigations was a modification of that applied to sunlight, differing from it chiefly in that it did not involve the use of a slit, and has since been generally followed in all cases in which it has not been essential to determine with a very high degree of accuracy the absolute— as contrasted with the relative—positions of the dark lines. A telescope is directed to the star, and the practically parallel rays falling upon the object-glass are by it condensed into a point-like image at its principal focus. In the ordinary use of the telescope these rays, continuing their courses, diverge again after meeting at the focal image, and, after traversing the eye-piece—a system of lenses equivalent to a magnifying-glass, enter the eye. It is convenient to regard the eye as directly observing the image at the focus of the object-glass by the aid of the magnifying eye-piece. If now a prism be placed in the path of the rays, either before or after they enter the telescope, since different colours are deflected differently, the point-like focal image becomes expanded into a line of coloured light. Such a line is, however, obviously inconvenient for examination; but by further interposing a "cylindrical lens"—a lens having for its faces portions of cylindrical instead of spherical surfaces—anywhere in the path of the rays, the line becomes broadened out into a band of definite width, in which dark lines are clearly visible. The line of light necessary for the purpose of analysis, instead of being obtained by a slit, is formed by the extension of the point-like image of the star into a line by the cylindrical lens. In Fraunhofer's arrangement of apparatus, the rays passed through the prism immediately before entering the telescope, and the cylindrical lens was so adjusted as to be just beyond the special line of light formed at the principal focus.

No observations sufficiently delicate to add anything material to Fraunhofer's discoveries in relation to stellar spectra were made before the publication of Kirchhoff's researches into the origin of the Fraunhofer lines in the solar spectrum. By these researches, the dark lines in the spectrum of the Sun were traced beyond doubt to the absorption of the colours corresponding to them in a solar atmosphere, and there could be little hesitation in extending the same principle to the explanation of the similar lines in the spectra of stars. The sun-like character of the stars, already apparent with respect to their total luminosity from the establishment of the Copernican system, became more intimately confirmed by the evidence revealed in the analysis of their light that, like the Sun, their glowing photospheres were enveloped in absorbent, and therefore cooler, atmospheres.

Although observations of the spectra of stars had been made for a few years previously by Father Secchi at Rome, their examination was first attacked systematically by Sir William Huggins and Dr. Miller about the year 1863. At that time, and even at the present, the visual observation of all but the most brilliant stellar spectra was a most trying and delicate task. The faint light of a star, even when collected over the extended area of a large object-glass and condensed to its focal point, is immeasurably feeble when compared with sunlight. It is necessary to further enfeeble it; first, by extending it into a spectral line, and, secondly, by expanding this line into a band; while, from the atmospheric unsteadiness, of which the familiar appearance of twinkling is a result, the excessively faint and, in most cases, barely visible spectrum is, together with its delicate system of lines, thrown into a continual state of tremor. In spite of such difficulties, however, Huggins and Miller succeeded in detecting in the spectra of several stars lines corresponding with those in the spectrum of the Sun as well as with many bright lines in the spectra of glowing vapours of terrestrial elements. They also thoroughly confirmed Fraunhofer's observations that in many cases the spectra of stars were strikingly different in the arrangement and intensity of their dark lines from that of the Sun.

While engaged in these observations, Sir William Huggins applied the spectroscope to the investigation of the physical condition of the nebulæ. We have traced in an earlier chapter the development of astronomical discovery and thought with reference to these cosmic clouds. We have seen that, at the time of Sir William Huggins's observation, the view was very generally entertained that they were stellar systems, the constituent stars being, in the great majority of cases, too faint to be individually distinguishable, though the great telescope of the Earl of Rosse appeared to have recently effected the resolution of some of the nobler examples. We have also followed the main features of Sir William Huggins's discovery. The telescope was directed to a small but rather bright nebula in the constellation of the Dragon. The image of the nebula, formed by the condensation of its rays by the object-glass, was no longer a point, as with a star, but an assemblage of points, one corresponding to each of the luminous points of the nebula; in fact, an image of the nebula, such that, if a screen or photographic plate had been placed behind the object-glass, at a distance from it equal to its focal length, a perfect, though excessively faint, picture of the nebula would have been formed upon it. As a cylindrical lens would not have expanded such an image into a line, it was considered expedient to make use of the more usual slit. The eye-piece of the telescope was removed, and a spectroscope was fitted in its place, the length of the collimator lying along the main axis of the telescope, while the slit was adjusted to the position of the principal focus of the object-glass. In this position the image of the nebula was formed upon the "slit-plate"—the pair of metal plates, by the separation of which the slit was produced. A thin slice of the nebula light thus entered the slit, and was subjected to analysis and subsequent examination in the ordinary way.

At a first glance the spectrum of the nebula appeared to be entirely monochromatic, its light being all condensed in a single green line. Closer examination, however, revealed the presence of a far fainter line rather higher up the spectrum—that is, toward the blue—as well as a third, exceedingly faint, and still higher in the spectrum. The failure up to that time to observe a spectrum of isolated bright lines otherwise than from the radiations of a glowing gas, had caused such a spectrum to be regarded as a crucial test of gaseous constitution—a conclusion thoroughly supported by all later spectroscopic work—and the observation was therefore universally accepted as demonstrating the gaseous constitution of the nebula. Of the three lines observed in the spectrum, the third coincided in position with the green line of hydrogen, the persistent character of which was so strikingly illustrated by Sir Edward Frankland and Sir Norman Lockyer five years later; neither of the other two probably correspond with any lines that have been obtained in terrestrial experiments, though, at the time, the first was thought to occupy the position of a line of nitrogen, to which, however, later measurements have shown it to be only exceedingly close.

Of the many nebulæ that have been subjected to spectroscopic examination since the date of Sir William Huggins's first observation, about one-half have been found to yield a spectrum of bright lines. It may be confidently asserted that the incandescent matter of all these is gaseous. With a few exceptions, the remainder yield faint continuous spectra unmarked by any evidence of special radiation or absorption, but it is not possible to infer from this alone that they are not either partially or entirely gaseous. Although a gas alone appears to possess the power of giving rise to a spectrum of bright lines, yet, under not abnormal conditions, its light may yield a continuous spectrum indistinguishable from that of an incandescent solid or liquid body. Excessive pressure and great depth of the radiating gas tend to bring about such a result, but a continuous spectrum has been obtained from a small quantity of oxygen contained in a glass tube under considerably less than the atmospheric pressure. It is not at present possible to interpret a continuous nebular spectrum.

The first and brightest of the nebular lines detected by Sir William Huggins appears to be specially characteristic of bright-line nebulæ. In every one of their spectra it appears as the brightest line of the series, while in many it occurs as the sole representative, other lines, though probably present, being too faint for detection. In a few instances other lines than the three originally seen have been observed, nearly thirty having been detected by visual and photographic observations in that of the Great Nebula of Orion. Of these, hydrogen lines, including the crimson c, and a yellow line due to the element helium, the significance of which will be seen later, are the only ones that have been reproduced in laboratory experiments, so that the chemistry of the nebulæ has only so far been connected with that of the earth through the elements hydrogen and helium.

In its first application the spectroscope had been essentially an instrument of chemical research. In its demonstration of the physical condition of nebulæ it had been brought to bear upon problems possessing a physical, as well as a purely chemical, interest; but it was now to invade the domain of physical science, pure and simple, and with the most remarkable and far-reaching results.

In 1848 Christian Doppler of Prague had directed attention to the fact that the apparent pitch of a musical note became affected during any variation in the distance separating the instrument emitting it from the ear. A note appears to rise in pitch as the source of sound approaches, and to fall in pitch as it recedes from the ear. The effect was recognized as a natural consequence of the wave transmission of sound, and Doppler showed, that, if light were also transmitted by wave motion, it should follow from analogous reasoning that the colour of an object should be affected by the motion of the source, becoming more violet as the object approached, and inclining toward red as it receded from, the observer.

It is a well-known fact that the sensation of sound

is due to the transmission of vibrations from a sounding body to the ear through the agency of wave motion in the air; and that the pitch of a note is the result of the frequency of the vibrations — the number executed in a second of time — a doubling of frequency causing a rise in pitch recognized by the ear as an octave. At each vibration of the sounding body a single wave is generated in the immediately surrounding air; the wave expands outward in an ever-increasing spherical surface—as a ripple on the surface of a pool unruffled by wind extends in a continually expanding circle—and, upon arriving at the ear, imparts to the auditory apparatus a vibration similar to that by which it originated. The vibration of the sounding body continuing, waves are continually generated, and follow in regular succession, so that, under normal conditions, as many enter the ear every second as are generated by the sounding body. The frequency of the note heard is therefore that of the source of sound. If, however, the source is approaching the ear, this correspondence is no longer maintained. The source generates waves with the same rapidity as before, a single wave being produced by each vibration; but it is important to notice that the speed with which the waves travel through the air—that is, the velocity of sound—is the same as when the source was at rest, since it may be shown from mechanical principles that the velocity of wave motion is determined solely by the physical properties of the medium in which they exist, and is entirely independent of any motion of the source. Since, therefore, the waves are travelling through the air with the same speed as when the source was at rest, and the source is now following them, they will be crowded together, and the length of each will be decreased. The waves, being shorter than before, and still forming a continuous series travelling with the original speed, will enter the ear in more rapid succession, the frequency of vibration of the auditory apparatus will increase, and the note will rise in pitch. From analogous reasoning it will be seen without difficulty that a recession of the source will result in a diminution of frequency of vibration, and consequently in a fall of pitch.

It may be well to give a further and more detailed illustration of this very important principle. Let us imagine a tuning-fork at rest, and radiating waves in the air, each 1 foot in length — waves that would correspond to a note about two octaves above the middle C of the piano. By the time ten vibrations had occurred, and ten waves had consequently been generated, the first wave would have travelled 10 feet from the fork, since the whole ten, each a foot in length, would form a continuous series. Now imagine the fork to move forward, and assume its velocity to be one-fifth that of the waves—that is, one-fifth the speed of sound—and under these new conditions imagine the original ten vibrations to be repeated. As before, ten waves will be generated, each following the last in regular succession. At the instant of generation of the tenth, the first wave will have reached the same point as before, its speed being unaffected by the motion of the fork; but, during its progress, the fork has been following it with a speed one-fifth of its own, so that the distance separating it from the fork is four-fifths of its former value. As there are the same number of waves lying between it and the fork, each must therefore be four-fifths as long as originally. The shorter waves, travelling with the same speed as before, enter the ear in more rapid succession, and the frequency of vibration is increased to five-fourths of its former value. In every case the change in frequency can be calculated in this simple manner from the relation between the speed of the moving source and that of the waves.

It is quite easy to notice the fall in pitch in the note of the whistle of an engine when passing the observer at express speed. For a speed of 60 miles an hour (88 feet per second), the speed of sound being 1100 feet per second, the reader should find little difficulty in showing that the frequency of the whistle is raised to 1·087 of its normal value while approaching, and decreased to ·926 of it while receding. The relative frequencies of 1·087 and ·926 correspond to an interval of nearly three semi-tones, that from *do* to *la*$_1$ in music, and such a change, which occurs at the instant that the engine passes the observer, can scarcely escape the notice of the least musical ear. The change is, of course, still more strongly marked when the observer, instead of being at rest, is travelling at express speed in the direction opposite to that of the whistling engine. The fall in pitch of the bell of a passing bicycle is quite appreciable to anyone with a fairly sensitive ear, even when at rest by the side of the road.

It is clear that if light depends upon the transmission of waves in the ether, similar changes must be produced in those waves by the motion of the source of light towards or from the observer. By a motion of approach, the waves entering the eye must be reduced in length; they must arrive in more rapid succession, and a colour more inclining to violet must result; while, from a motion of recession, the waves must be drawn out; they must enter the eye in less rapid succession, and the colour must appear lower in the spectral series. Such was the prediction of Doppler, and he suggested that the strongly-marked colours of certain stars might originate in their rapid motions.

At the time of the enunciation of this principle, the most serious objection to its suggested application appeared to lie in the excessive speeds with which it was necessary to suppose the stars to be endued. By the rush of a star towards the Earth, the whole series of spectral colours present in its radiations were supposed to move up the spectrum, so that the light received appeared to be abundantly rich in violet and poor in red rays, the intermediate colours appearing the same as before, since, as each became displaced towards the violet, its place would be supplied by the similarly affected rays immediately below it. Owing to the high velocity of light, it was clear, that, to effect any appreciable change of colour by such a process, velocities of many thousands of miles per second must be imagined among the stars, and there appeared to be no warrant for so extravagant an assumption.

A still more fatal objection to Doppler's theory became apparent, when, in the years immediately following, the extension of the spectrum into its invisible ultra-violet and infra-red regions was discovered. From the existence of these invisible radiations, it would follow that the colours of the visible spectrum of a star should be unchanged by its approach or recession. A motion of approach would cause the frequency of each radiation to increase, and, in consequence, each colour would become more violet in hue, and would experience greater deviation by the prism. Each colour would, therefore, be displaced towards the violet end of the spectrum, assuming the previous tint, while occupying the former position, of the colour immediately before it. The lowest red rays would move farther into the spectrum, their colour at the same time becoming brighter; but the frequency of the rays immediately below them, previously just too low to excite the sensation of vision, would be so increased that they would appear in the spectrum just within its lowest limits, and would take the place of those deep rays that had been displaced upwards. Similarly, the frequency of the extreme violet waves would be so increased that they would enter the invisible region beyond. The colours of the visible spectrum would therefore be the same as when the star was at rest.

There is no doubt that the colour of a star is a physical fact, initially impressed upon its radiations, and that it cannot be explained by any theory of optical illusion.

So far the attempt to apply Doppler's principle to physical astronomy had been attended with failure; but in 1848 Fizeau indicated a method by which it might still be possible to detect by its aid evidence of the approach or recession of stars in their analysed light. It was suggested that, instead of attempting to detect the evidence of motion in the entire light of stars, close attention should be directed to the exact positions of the dark lines by which their spectra were crossed. We have seen that a dark line in a spectrum indicates the position in it of colour absent from the radiations; and it follows that, by a motion of approach of a star, since the colours that are just more and just less refrangible than the missing one will be equally displaced up the spectrum, the gap separating them will be similarly displaced; in other words, every dark line in the spectrum of an approaching star should be displaced toward the violet of the spectrum, while, from analogous reasoning, every dark line in the spectrum of a receding star should be displaced toward the red.

After several unsuccessful experiments, Sir William Huggins felt justified in announcing in 1868 that he had succeeded in detecting in the spectrum of Sirius such a displacement of one of its spectral lines as would result from a motion of recession of the star. The selection of Sirius for the purpose of the research was due to the great intensity of its light, as well as from the strongly-marked character and undoubted origin of its spectral lines. It has been found, however, more recently, that these advantages are seriously discounted by the grave disadvantage arising from the ill-defined character of the lines. The visual spectrum of Sirius differs from that of the Sun in the extraordinary emphasis of the dark lines of hydrogen absorption—indeed, under ordinary conditions these are all that are visible, though, when the atmosphere is steady and an instrument of high optical quality is employed, a great number of fine dark lines, many of them corresponding with bright lines in the spectrum of iron, may also be distinguished. The dark hydrogen lines in the spectrum of Sirius are about six times as broad as those in the solar spectrum, and, unlike the latter, which are clearly marked and sharply defined, those of the star are hazy and pass by insensible degrees into the bordering light of the spectrum. There is good reason to regard this difference as indicating a greater density of the hydrogen in the atmosphere of the star; since, while at a low density, as in the ordinary vacuum-tube, glowing hydrogen gives a spectrum consisting of bright lines that resemble the dark lines in the spectrum of the Sun in being fine and sharp, with increased density of the gas the lines become broad and badly defined at their edges; and when the gas is under a pressure approaching, though still distinctly below, that of the atmosphere, its bright lines very closely resemble in their definition the dark lines of the Sirian spectrum.

To determine whether a dark line in the spectrum of a star coincides exactly with a corresponding bright line in the spectrum of an incandescent terrestrial vapour, it is absolutely necessary that both should appear in the field of view at the same time, and this condition necessitates the rejection of the more convenient cylindrical lens in favour of the slit. In Huggins's experiment the spectroscope was so adjusted that the slit was very near but not coincident with the principal focus of the object-glass of a telescope of 8 inches of aperture, so that a small length of it became illuminated by the nearly condensed rays of the star, and a spectrum of a corresponding width was produced. At the same time, rarefied hydrogen, contained in a glass tube placed just beyond the object-glass, was made to glow by the electric discharge from an induction-coil. Under these conditions the narrow spectrum of the star became visible, while extending right across it were the bright lines of the glowing hydrogen. Attention was specially directed to the most conspicuous of the dark lines—that in the green part of the spectrum, the far more delicate bright-green line of the glowing gas appearing to traverse it in the direction of its length. The most careful observations of the lines were made many times, and Huggins felt confident that the bright line, though it appeared projected upon the dark one, did not lie along the middle of it, but was placed rather lower down toward the red of the spectrum. Since, however, it was quite conceivable that expansion of the dark line in the spectrum of the star, presumably due to increased density of the absorbing gas, might not have taken place equally upon both sides, the observation was not quite conclusive, but Huggins carefully examined the spectrum of glowing hydrogen at varying densities, and found that the broadening accompanying increase of density was in every case entirely symmetrical upon either side. The want of coincidence between the centres of the lines was therefore confidently attributed to motion in the line of sight, and, from its extent, after allowing for the motion of the Earth in its orbit at the time, Huggins estimated that the star was receding from the Sun at the rate of $29\frac{1}{2}$ miles per second.

In the following year Huggins extended his observations to other stars, and was successful in detecting displacements of spectral lines in thirty instances. With some stars, such as Rigel and Castor, the hydrogen lines were displaced towards the red end of the spectrum, an indication of recession; with others, including Arcturus and Vega, they were raised towards the violet, and denoted approach. Further, with the view of confirming beyond doubt both the soundness of the principle and the possibility of its practical application, the spectrum of Venus was observed at times when, from the position of the planet in its orbit, its motion was known to be directed towards and away from the Earth. Since the light of Venus is reflected sunlight, the ordinary Fraunhofer lines are represented in its spectrum, but a careful examination of the selected hydrogen lines showed, as had been confidently anticipated, displacements from the positions of the lines of terrestrial hydrogen, and precisely such displacements as were demanded by theory from the speed of the planet relatively to that of light.

The approach or recession of an object is known technically as its motion in the line of sight. It is clearly but one component of the whole movement, the other being a drift at right angles to or athwart the line of sight, and the determination of the complete motion demands a knowledge of both of the components. Motion of a star across the line of sight is indicated in its so-called "proper motion", or apparent rapidity of drift across the face of the sky, but, since the apparent drift represented by a given velocity obviously depends inversely upon the distance of the star, it is essential to know the distance before translating proper motion into definite velocity. Our present knowledge of the distances of stars is so imperfect that only in a very few instances is it possible to make the application with any approach to accuracy, but in a few instances some rough approximation has probably been possible. The unique power of the spectroscopic method of determining motion lies in the fact that the exact interpretation of its record is entirely independent of the distance of a star, the same displacement of spectral lines resulting from a definite movement in the line of sight, whether the luminous body is a member of the Solar System,

The Analysis of Starlight.

or whether it lies at the extreme limits of fathomable space.

There can assuredly be but little hesitation in placing the detection and measurement of the motions of stars in the line of sight as among the greatest achievements of physical science. Before the statement of Doppler's principle, the mere detection of such movements must have appeared to be beyond the very possibility of human endeavour, at any rate without observations extending over many thousands of years. From the motion of a star in line of sight, its position upon the face of the heavens is unaffected. It is true that by the continuance of such movement, in the course of time a change in the apparent brightness of a star would result, as also an alteration of its parallax; but so many thousands of years must elapse before either would become appreciable to the most refined observation, that but little enthusiasm could be aroused by the contemplation of the ultimate possibility of the successful application of either method. By the discovery of Huggins, however, an observation, demanding, it is true, the utmost delicacy, but which need not extend over more than a few minutes, has proved sufficient. Again, few discoveries have furnished a finer illustration of the debt that one science so frequently owes to its sisters, and of the unexpected nature of the conjunction towards which accumulation of knowledge is tending. Upon one line we see the story of sunlight first roughly sketched in the ray dispersed by Newton's prism; told with more detail in the improved conditions of experiment devised by Wollaston and Fraunhofer; and again with a still fuller meaning in the shrewd conjectures of Stokes and in the experiments of Kirchhoff. Upon a converging line we trace the first conception of the wave theory of light in the genius of Huygens, and, after a century of neglect, its restoration and establishment upon a firm foundation by Young. Anon comes Doppler still discovering, though mistaken as to the exact course he was directing; then the direction of the course towards its proper goal by Fizeau; and, later, its magnificent attainment in the experimental skill of Sir William Huggins.

In 1870, two years after Huggins's discovery, Dr. Hermann Vogel, who then had charge of a private observatory at Bothcamp, was attracted to the measurement of stellar motions in the line of sight. For four years at Bothcamp, and for a further period of thirteen years at Potsdam, where he became possessed of instrumental means of greater power, Vogel carried out measurements upon practically the same method as that originally adopted by Huggins. In 1887, however, by which time the photographic gelatine dry plate—first introduced by Mr. Kennet in 1876—had reached a high degree of perfection, and had been introduced with remarkable success in other branches of astronomy, Vogel applied it to the purpose of his work, and soon became convinced, that, by photographing the spectrum of a star together with the bright lines of glowing hydrogen or some other terrestrial vapour introduced for the purpose of exact comparison, and by subjecting the compound spectrum so photographed to microscopic examination, a far higher degree of accuracy could be attained than was possible in visual observation. From that date to the year 1891 the photographic method was consistently followed at Potsdam, and from the consistency between independent observations of the same star at different times there can be no hesitation in regarding the results as constituting the most exact record that has so far been acquired of the motions of stars in the line of sight. The speeds of approach and recession of the following eight more familiar stars are taken from Vogel's results, the velocities being given in miles per second. Due allowance has been made in every instance for the direction of the speed of the Earth in its orbit—18·7 miles per second—at the time of observation, so that the numbers are actually the velocities of the stars relatively to the Sun. So far, no star has been found to possess a higher velocity in the line of sight than Aldebaran.

VELOCITIES OF APPROACH AND RECESSION OF STARS
(Miles per second).

Approaching Stars.		Receding Stars.	
Sirius,	9·8	Aldebaran,	30·2
The Pole Star,	16·1	Rigel,	10·2
Arcturus,	4·8	Capella,	15·2
Vega,	9·5	Betelgeux,	10·7

It was during the course of these observations that the last step was taken in the demonstration of the existence of the dark companion of Algol. The essential principles—other than the spectroscopic ones—that underlie the investigation have already been given in an earlier chapter,[1] and the reader will now have no difficulty in completing the story of the discovery. It had been shown to follow from the laws of motion that, if the regular and continually repeated fading observed in the light of Algol were due, as was strongly suspected, to its periodic eclipse by a dark star revolving round it in an orbit presented edgeways to the Earth, Algol itself should be in revolution in a similar orbit, and in the same plane, and should therefore during each revolution alternately approach and recede from the Earth. Such approach and recession should produce an oscillation of the dark lines in the spectrum of the star, as they became by Doppler's principle displaced higher and lower in the spectrum, and precisely such an oscillation of the lines as was demanded by theory was actually detected in twelve photographs of the spectrum of Algol taken at intervals between 1889 and 1891.

An oscillation of spectral lines precisely similar to that presented by Algol was discovered during the same period in Spica, the most brilliant star in the constellation of the Virgin. In this case the complete oscillation was effected in just over four days, and the orbital speed of the star indicated by the displacement of its spectral lines is 56·7 miles per second. It can, therefore, scarcely be doubted that, like Algol, Spica is accompanied by a dark companion, but that the plane of its orbit is inclined

---

[1] See pp. 22–31.

to the line of sight to such an extent that, at each conjunction, the dark star passes either just over or under it, thus avoiding an eclipse. Vogel is indeed of opinion that faint traces of the spectrum of the companion can be detected in the photographs.

While these refined observations were in progress at Potsdam, a very beautiful application of Doppler's principle was effected at the observatory of Harvard in the United States. From 1886 to 1890 the energy of Professor E. C. Pickering and his assistants was mainly directed to effecting a photographic record of the spectra of stars as far as the eighth magnitude. The method adopted consisted in accurately adjusting a photographic plate at the principal focus of the object-glass of a telescope, while immediately in front of the object-glass, a large prism—known as an objective prism—was fixed. In the absence of the prism the sensibly parallel rays of a star would have been condensed into a point-like image upon the plate, but by the prism the image was elongated into a spectral line. The method so far was, as we have seen, that devised by Fraunhofer, who expanded the spectral line into a band of sensible width by a cylindrical lens. In the work at Harvard, however, no cylindrical lens was employed. The line was directly photographed, but, by causing the telescope to slowly move relatively to the star so that the spectral line drifted in a direction at right angles to its length over the plate a band was produced in which the dark lines were distinctly visible. All stars included within a certain small area of the heavens toward which the instrument was directed formed images, and therefore spectra, upon the plate, which therefore frequently contained a considerable number of stellar spectra. The method was inferior to that followed at Potsdam in that, in the necessary absence of a comparison spectrum from the plate, it was impossible to determine the absolute positions of lines in their respective spectra with great accuracy, such, for instance, as would have been essential to the detection of motion in the line of sight, but it enabled a far greater number of stars to be examined in the time, the spectra of over ten thousand being in fact photographed and examined in four years. Upon examining several photographs of the spectrum of Mizar, the middle star of the three forming the handle of the "Plough" or the tail of the "Great Bear", the singular fact appeared that while upon some plates the dark lines presented a normal appearance, on others they were doubled; and upon a more critical examination it appeared that they opened and closed with perfect regularity in successive periods of fifty-two days. A simple and complete explanation of these appearances is found in the assumption that we are here presented with a system of two stars, similar to that of Algol and its companion, except that in the case of Mizar both of the stars are bright. The actual photographed spectrum would therefore be a combination of the spectra of the two stars. The pair being in continual revolution round their common centre of mass, at the instant of their conjunction with the direction of the Earth they would be drifting across the line of sight in opposite directions; neither would be approaching or receding; the spectral lines of both, being in their normal positions, would coincide; and the appearance of a single spectrum would result. A quarter of the complete period of revolution later, however, one of the stars would be rushing towards and the other from the Earth, their lines would consequently experience displacement in opposite directions in their spectra, and would appear as separated. In another quarter period half a revolution would have been accomplished from the time of the first observation, conjunction with the Earth would again occur, and the combined spectrum would once more assume its normal appearance. There can be little hesitation in accepting this explanation.

From the impracticability of determining the exact positions of the lines in the spectrum it was not possible to form an estimate of the actual speeds of the component stars in the line of sight, but, from the widest distance to which the lines open out it is a simple matter to determine for the instant of their greatest separation the speed of one star relatively to the other in the line of sight. It appeared to be about 100 miles per second. If we assume that the orbits are presented edgewise to the Earth, the movement of the stars at the time of the widest separation of their spectral lines would be directly towards and away from the Earth, and this velocity would be the actual speed of one star relatively to the other. From the knowledge of this relative velocity, and of the complete period of revolution—in this case 104 days—it is possible from the laws of mechanics and gravitation to calculate the combined mass of the stars. In the result a mass is indicated of forty times that of the Sun. If the orbits are merely inclined to the direction of the Earth and are not presented edgewise, the actual relative speeds, being at the time of greatest separation of the lines only in part directed to the Earth, must be greater than those assumed, and the masses of the stars must exceed the value deduced from the first assumptions. Since there are no means of determining the inclination of the orbits to the line of sight, it becomes therefore only possible to determine a limit above which the mass of the double star must lie.

From the spectroscopic examination of an object so remote that its distance is incapable of determination, and that in the field of view of the most powerful telescope appears but as an absolute point of light, to see a pair of revolving suns; to measure the period of their mutual revolution; to trace over their blazing surfaces cooler atmospheres, and in these to recognize gases familiar upon the surface of the Earth; and to assign a minimum limit to the mass of the entire system, is an achievement that can scarcely fail to appeal even to those many and most hardened of sinners against intellectual light who would value scientific investigation only in exact proportion to the monetary equivalent of its technical application.

Chapter VI.

The Red Flames of the Sun.

We have traced in a previous chapter the course of discovery resulting from the spectroscopic examination of the general light of the Sun. From the time of Fraunhofer's observations to those of Ångstrom, the light submitted to examination was indeed that of the Sun, but it is important to notice that no pains had been taken to differentiate the radiations thrown off by different parts of the Sun's surface. From the centre, as well as from the edge of the disc; from the dark spots and brilliant faculæ, as well as from the delicate extensions of its atmospheric surroundings, only so far recognized during the brief moments of totality of solar eclipse; rays entered the spectroscope and mingled their story in the resulting spectrum. In 1866, however, Ångstrom for the first time adopted a different method, one by which it became possible to examine in detail the radiations of different parts of the Sun's surface, and which has during the years that have followed yielded a veritable harvest of interesting and valuable results. It will only be possible in the present chapter to pass under review very briefly a few of the more remarkable of these in their bearing upon the Physics of the Sun.

Ångstrom's device consisted in forming, by means of a convex lens, a sharply-defined image or picture of the Sun upon the slit-plate of a spectroscope.

The result is generally and most conveniently obtained by adjusting a spectroscope so that its collimator lies along the axis of an astronomical telescope, and so that its slit-plate is removed from the object-glass by a distance equal to its focal length. Under these conditions, and when the telescope is directed towards the Sun, the rays from different portions of the Sun's surface are focussed at corresponding points upon the plate, and there is formed upon it a perfectly-defined picture of the solar disc, in which the sun-spots are clearly visible. The size of the image is, by elementary optical laws, in direct proportion to the focal length of the lens forming it, a picture of the Sun an inch in diameter necessitating a focal length of about 9 feet. To observe the spectrum of any selected portion of the Sun's surface, the slit is so adjusted that the image of that particular portion falls upon it. Under these conditions, the converged rays from the selected region of the Sun, instead of illuminating the plate, pass directly through the opening of the slit, and travelling through the spectroscope, are subjected to analysis. It will be noticed that the method is essentially the same as that applied by Huggins two years earlier to the examination of the spectra of nebulæ. It had also been employed by Donati in 1864 for the purpose of examining the spectrum of a comet.

It is important to observe that different portions of the length of the slit being illuminated by different parts of the Sun's surface, the light filling it may, and indeed frequently does, differ in quality in different parts of its length. As in the resulting spectrum the light from each small part of the slit is spread out into a spectral band, the appearance is produced of the several and frequently differing spectra of those portions of the Sun's surface by which the slit is illuminated arranged as a series of parallel strips, each being in contact with those immediately above and below it.

In 1868 the new method was applied to the study of the "red flames" or "prominences" of the Sun. At a total eclipse of the Sun, during the few minutes at most in which the glowing photosphere is covered by the Moon, there are seen, apparently projecting from behind the dark disc of the Moon, the delicate appendages of the Sun known as the "prominences" and the "corona". The corona appears as an exquisitely beautiful and generally irregular halo of silvery light surrounding the black circle of the Moon. It is full of the most delicate detail, and appears to the eye to consist chiefly of streamers, some of which frequently extend from the surface of the Sun to a distance greater than its diameter. The far smaller but more brilliant "prominences" are rose-tinted projections that frequently assume the most fantastic forms, and occasionally extend to a height of a quarter of the Sun's diameter from its surface. Though at one time generally regarded as belonging to the Moon, the prominences had for some years been recognized as of solar origin, from the fact that during the progress of an eclipse, the dark disc of the Moon had been seen to travel over them, gradually covering those in front, and at the same time unveiling those behind the direction of its motion. A sufficient explanation of their invisibility upon the limb of the uneclipsed Sun, even when viewed with the telescope, is that their fainter light is entirely overwhelmed by the far more brilliant illumination of the Earth's atmosphere produced by the direct solar rays.

In 1868 there occurred an eclipse of the Sun in which the track of the Moon's shadow travelled across India. It was observed from several stations situated at different points of the shadow's path by scientific men collected from all parts of the civilized world, and among them the French astronomer M. Janssen, who had made special arrangements to examine the spectrum of the prominences. As they flashed out at the instant of totality, Janssen rapidly brought the slit of his spectroscope across the telescopic image of one of the finest, and, on applying his eye to the instrument, perceived at once a number of bright separated lines, the certain indication of gaseous constitution. But this was not all. Janssen was so impressed with the extreme brightness of the lines, that, as the prominences themselves melted from view in the reappearing sunlight, he perceived the possibility of recognizing them on the edge of the uneclipsed Sun. Clouds prevented him from attempting the observation during the remainder of that day, but on the following morning, not long after sunrise, Janssen carefully searched the immediate neighbourhood of the Sun's limb with the spectroscope, and had no difficulty in again recognizing the brilliant lines of the spectra of prominences entirely invisible in the telescopic view of the Sun.

The principle underlying this discovery, a discovery that initiated an entirely new application of the spectroscope, is extremely beautiful, as well as simple. The atmospheric glare in the direction of the Sun, by which the prominences are usually concealed, is scattered sunlight, and as such, yields the ordinary solar spectrum. The amount of light distributed over the whole of this spectrum is derived from that entering the slit, so that the intensity of illumination, or brightness, of the spectrum becomes less and less in direct proportion to its extension in length, and must continue to do so until the dispersion becomes so great that the spectrum, ceasing to be continuous, breaks up into a number of separate images of the slit. We have seen, however, that with sunlight this has never been effected. Since, by the employment of a number of prisms, through all of which the light is transmitted in succession, any required degree of spectral extension may be obtained, the brightness of the spectrum may by the same means be reduced to any required extent. With the prominences, however, this is not the case. From the fact of their being gaseous, their light consists of a finite number of, and practically of a very few, pure colours. The first separation of these in the spectroscope is, of course, accompanied by a decrease in brightness, but since the colours are now pure, they are not further enfeebled to whatever extent the spectrum is extended; the only effect of such extension being to place them farther apart. We have then the general result, that the brightness of the spectrum of the prominences is only slightly, and to a limited degree, enfeebled by spectral extension; whereas the spectrum of the atmospheric glare is enfeebled without limit in direct proportion to spectral extension or dispersion. After a certain dispersion, therefore, the general illumination of the greatly weakened spectrum of the air glare is no longer sufficiently bright to conceal the scarcely reduced light of the spectrum of the prominences, and their bright lines become visible upon the background of the enfeebled spectrum of the sunlight scattered in the Earth's atmosphere.

Previously to Janssen's discovery, however, the principle by which it was effected had been clearly recognized by the English astronomers. It had been plainly stated in 1866 by Sir Norman Lockyer, who, at the time of the eclipse, had a powerful spectroscope in process of construction for the purpose of attempting its application. The instrument was completed shortly afterwards; and, by its means, Lockyer detected the spectra of prominences before the news of Janssen's success had reached England. Sir William Huggins had, moreover, already searched the limb of the Sun for spectra of prominences with a spectroscope of moderate power, though without success. Upon the announcement of the discovery, however, he repeated the observations with the same instrument; and, now that he was aware of the exact part of the spectrum toward which to direct his attention, had no difficulty in recognizing the bright lines.

The bright lines of the spectra of prominences had been observed by several astronomers in India during the progress of the eclipse. Although other fainter lines had appeared, one observer having indeed detected as many as nine, by far the most conspicuous were three—a crimson, a green, and a yellow. The duration of the eclipse had been too brief to allow of an accurate determination of the positions of the lines, but it was believed from their general appearance that the red and green would be found to be due to the radiations of hydrogen, and the yellow to those of sodium. On subsequently examining the spectra of prominences at leisure in the uneclipsed Sun, the coincidences of the red and green lines with those of glowing hydrogen were established, but the yellow line was found to be rather more refrangible than the yellow of sodium, and not to correspond in position with any line that had so far been recognized in the spectrum of a terrestrial gas. It was consequently assumed to arise from the radiations of a gaseous constituent of the prominences unfamiliar to terrestrial chemistry. Later, this hypothetical gas received, at the suggestion of Frankland, the name of helium. For a long time subsequently, helium remained unrecognized, save as a constituent of the Sun, stars, and nebulæ, until, in 1895, it was discovered by Professor William Ramsay among the gases extracted from a terrestrial mineral, clevite.

For some days after the eclipse Janssen remained at his station in India, fascinated with the application of the new discovery. He soon found that the bright lines originally seen in the spectra of prominences could be traced, though to a far less distance from it, round the entire limb of the Sun. There could be no hesitation in ascribing their continual appearance to the radiations of the incandescent atmosphere of the Sun, and it consequently appeared that the prominences were essentially enormous and local extensions of the solar atmosphere. Janssen also found that the prominences were essentially unstable, and that they were subject to changes upon the most stupendous scale. During the eclipse an enormous red flame, estimated as being at least 89,000 miles in height, had been directly seen upon the edge of the Sun: but, upon the following day, scarcely a trace of bright lines could be detected by the spectroscope in the place that it had occupied. From day to day Janssen traced, from the occurrence and the varying distance to which they could be followed from its limb, the mighty surging now recognized for the first time in the atmosphere of the Sun.

In this, the earliest method of studying prominences upon the limb of the uneclipsed Sun, observation was confined to a narrow section of the prominence the image of which was at that instant formed upon the slit of the spectroscope. It was possible, however, from the examination of a number of such sections, to trace the complete form of the prominence. For this purpose it was most convenient to adjust the slit so that it lay "radially", or at right angles to the edge of the solar image formed upon the slit plate, and so that a small portion of it penetrated the image itself. The appearance in the spectroscope then consisted of the ordinary solar spectrum of that portion of its surface that illuminated part of the slit, while above it was extended the bright-line spectrum of the solar atmosphere and prominences, the length of the lines depending upon the extension of the radiating gases above the solar surface. By moving the spectroscope so that the slit travelled round the solar image while always maintaining its radial position with reference to it, the bright lines were seen to contract or extend as the level of the atmosphere and prominences rose and fell. From measurements of the lengths of the lines, it therefore became possible to trace the varying height of the incandescent gases producing them, and thus to construct the complete outline of a prominence.

In the early part of the following year a further simplicity was effected in the observation of prominences. In 1869 Sir William Huggins, having by its spectrum detected a prominence upon the limb of the Sun, carefully adjusted the slit so as to lie within the image of the prominence but just outside of that of the Sun. He then boldly opened the slit, and saw, not the line spectrum of the prominence, but, in the position where its red line had appeared, the prominence itself, splendidly displayed as a cloud of crimson fire.

The explanation of this very beautiful discovery is again extremely simple. The opened slit may be regarded as a narrow window directed towards the prominence. The light of the sky entering the window along with that of the prominence, is spread out into a very impure spectrum, which, retaining its continuous character, is greatly enfeebled in brightness by its extension. The prominence light, however, consisting almost entirely of three pure colours, is at once resolved, but is not further weakened. The crimson rays of the prominence are deflected in block by the prism, and, being scarcely reduced in intensity, convey to the eye the appearance of a crimson prominence superposed upon the enfeebled crimson of the spectrum of the atmospheric glare. In a similar manner, a green picture of the prominence is produced by its green rays, and appears upon the green region of the air spectrum, while each pure colour present in the radiations of the prominences must, in like manner, develop a picture of the prominence in its own colour. Until very recently this constituted the most powerful method for studying the varying forms of the prominences. From the great visual intensity of the crimson light, the picture formed by it has been the one generally selected for observation.

The picture presented by solar prominences, when viewed with fine instrumental means, is of such extreme beauty as to lend a special charm to their study. As was first indicated by Lockyer in 1870, they appear to be divisible into two distinct classes. Those of the first, which are generally known as *quiescent*, are commonly the larger. In appearance they closely resemble terrestrial clouds bathed in the crimson glory of sunset. Their soft outlines experiencing but gradual change, they seem to float, often for days together, at a great elevation above the glowing surface of the Sun, being sometimes connected with it by delicate filaments of light, but at other times entirely separated from it. They are to be traced round the entire limb of the Sun, and appear to have no direct relation to sun-spots. The spectroscope shows their light to consist almost entirely of the radiations of hydrogen and helium.

The prominences of the second class, which are known as *eruptive*, appear, as their name implies, to be the results of veritable explosions from beneath the cloud-surface of the photosphere. They are intensely brilliant. They are subject to the most violent and rapid changes, and are short-lived. Their spectra indicate, that, while their light is, in the main, due to the glowing hydrogen and helium, the vapours of other metals, and conspicuously those of sodium, magnesium, and iron, are generally present in them. They are clearly connected with the unrecognized physical cause to which sun-spots are due, for they are most densely distributed over the zones on either side of the Sun's equator to which spots are almost entirely limited; they appear in greatest abundance at the approximately regular periods of eleven years that mark the maximum richness of spot distribution; and, in several instances, they have been seen in direct connection with large spots, that have, by the slow rotation of the Sun, just reached the edge of the disc. When it has been possible to trace the connection most exactly, their glowing matter has appeared to have been projected from the edges of spots, and later, in its descent, to have fallen towards their dark and probably depressed centres.

Violent commotions, evidenced by rapid changes in appearance, are invariably associated with prominences of the eruptive class, but they have been, from the first days of prominence study, recognized in another manner. The spectrum of a prominence is usually observed with the slit, which is now very narrow, lying radially across the limb of the Sun's image. With such an arrangement the bright lines of the prominence spectrum seem, under normal conditions, to be the continuations of the corresponding dark Fraunhofer lines in the solar spectrum appearing immediately below and in contact with them. Frequently, however, this coincidence is not maintained, a prominence line appearing to be displaced to one side or the other of the position of the corresponding Fraunhofer line, while occasionally displacements in different directions are observed as the slit is adjusted to different parts of the image of the same prominence. The reader will have no hesitation in assigning these displacements to their true cause—to a Doppler effect, due to the rush of glowing prominence matter either directly from or towards the observer. In this manner velocities in the line of sight have been recognized of from 200 to 300 miles a second, a velocity upon much the same scale as that traced across the direction of vision by the rapidity of changes in prominence forms. From the detailed method of observation, however, spectral changes, due to movement in the line of sight, are now more complicated and fuller of meaning. Since different parts of the slit are illuminated by light from different regions of a prominence, and as each element of length of the slit gives rise to a correspondingly narrow strip of the spectrum, it frequently happens that, in different parts of its length, a bright line displays different displacements, due to differing velocities in the line of sight in the corresponding regions of the prominence. Thus, one part of a line may be displaced to the red, while another is displaced to the violet, and the whole line not unfrequently assumes a curiously curved and irregular form. It is clear that, from the study of these contorted lines, the motion of the glowing matter in the line of sight may be determined at different levels of the prominence.

For twenty-two years the method discovered by Huggins for observing prominences remained supreme. In 1891, however, Professor Hale of Chicago devised an extremely beautiful instrument, which he has termed the spectro-heliograph, by which it has now become possible to record their pictures by photography in a far more expeditious and perfect manner. The principle of the spectro-heliograph is as follows. Although, in the total quantity of their light, the prominences are so much less brilliant than the glare produced by the Sun's rays in the atmosphere of the Earth, that they are, excepting when the Sun is totally eclipsed, permanently invisible; yet, in their special radiations, they are so much brighter, that, when observed under such conditions that attention is only directed to these, they become clearly visible. Professor Hale's method consists essentially in photographing the Sun and its immediate surroundings by light of one colour only, selecting as that colour one especially strongly represented in the radiations of prominences. The prominence ray that appeals most powerfully to the eye is, as we have seen, the crimson light of hydrogen; but this is quite unsuitable for the purpose of photography, since it produces no effect upon the ordinary photographic plate, and those plates that are specially prepared so as to be affected by it are very slow in their action. The more refrangible green light of hydrogen might conceivably be employed; but, still better than either of these, are either of two deep violet rays present in all prominence radiations, and corresponding in their positions in the spectrum with the two broad dark Fraunhofer bands H and K that lie almost at the violet extremity of the visible spectrum. Owing to the extreme position of these rays in the spectrum they scarcely affect the eye, and for this reason would be entirely unsuitable for visual observation, but they are extremely energetic in their action upon the photographic plate. Their radiations are, as we have already seen, due to the incandescent vapour of calcium. Although always present in the light of prominences, owing to the feebleness of their visual effect they remained undiscovered until 1882, when they were recognized in photographs of the spectra of prominences seen upon the edge of the dark moon during the famous total solar eclipse of that year. Although the dark bands H and K are very broad, the corresponding violet lines of the prominence spectra are fine; hence there is this further advantage in making use of them, that, since the colours immediately next them in the spectrum of sunlight are absent or weakly represented, they are more easily isolated for the purposes of experiment. In practical work it has been found most convenient to make use of the violet line that corresponds with K.

In photographing prominences with the spectro-heliograph, an image of the Sun is formed by the object-glass of a telescope upon the slit plate of a spectroscope in the usual manner, the image being, in the apparatus actually employed, about 2 inches in diameter. The extremely narrow slit, which is somewhat longer than this, lies right across the picture, and is therefore illuminated by light from a narrow strip of the Sun and its atmospheric surroundings on either side. The spectrum is focussed upon a small screen which forms a part of the instrument, and it of course consists of a succession of images of the slit, one formed by each colour present in the light that penetrates it. In the screen there is a second slit, and this is carefully adjusted so as to be in the exact position of the bright K line. The violet K light, therefore, and that only, penetrates the second slit, and forms an image of the first upon a photographic plate that is placed immediately beyond. There being formed upon the surface of the plate an exact picture of the illuminated slit, so far as its K radiation is concerned, the picture of the slice of the Sun, the image of which is focussed upon the slit, is therefore reproduced by these rays. By the regular action of a water-clock, the slit of the spectroscope is now made to travel across the image of the Sun, thus successively including images of all portions of its surface, while, by a proper mechanism, at the same time, and with such a perfectly sympathetic movement that the K line of the moving spectrum continually coincides in position with the second slit, the screen in front of the photographic plate is carried over it. Adjacent pictures of adjacent strips of the Sun are therefore photographed by K light, and thus, in the result, a complete picture of the Sun is produced, while the prominences are so rich in K rays that they appear beautifully defined upon it. The passage of the slit over the whole image of the Sun occupies but a fraction of a minute, in which time a survey is effected that would occupy an observer several hours to complete less perfectly by visual observation.

Professor Hale's method is invaluable in recording not only prominences, but other features of the solar surface to which we have not as yet referred. Near the limb of the Sun, and most richly displayed in the neighbourhood of sun-spots, there are always visible in telescopic observation irregular bright masses, commonly forming a rough network over the surface. The visibility of these masses, which have received the name of "faculæ", near the limb, combined with the fact of their disappearance as they are carried by the rotation of the Sun farther on to its disc, is satisfactorily explained by the assumption that they are not appreciably brighter than the clouds of the photosphere, but that they float at a higher level. The brightness of the solar disc is readily seen through a telescope to diminish towards the limb, a consequence of absorption exercised upon its light by its atmosphere, the absorption being specially pronounced in rays coming from the limb to the eye, by reason of their oblique passage through the atmosphere, and the consequent great length of their path involved in it. The faculæ being at a higher level than the general surface, their light does not experience the effects of absorption in so marked a manner, and when near the limb they therefore become visible upon the background of the dimmed photosphere.

In 1872 Professor Young of New Jersey observed, that, in the spectra of faculæ, fine bright lines always appeared down the centres of the broad bands H and K. The appearance probably indicates that the faculæ contain the incandescent vapour of calcium at a lower density but at a higher temperature than the same vapour that, in the atmosphere at a lower level, produces by its absorption of the light of the photosphere the bands H and K. According to this view, the light of the photosphere is robbed of the H and K radiations while traversing the cool and dense mass of calcium vapour lying immediately above it; at a still greater height more intensely heated clouds of the same vapour in part restore the rays, but the glowing matter being now at a lower density, the light is more truly monochromatic, and narrower spectral lines are the result. If, therefore, it were possible to view the Sun by its K light, and that alone, we should in all probability be able to distinguish the faculæ not only near the limb but over the whole disc; and the method applied by Janssen to the prominences would probably be successful but for the fact that these extreme violet radiations affect the eye to so slight an extent. The photographs of the spectro-heliograph are, however, entirely taken by K radiation, and it is not therefore surprising to find in them representations of faculæ over the entire picture of the Sun. It should be added, that when it is desired to photograph the faculæ the operation is carried out precisely in the manner described, but that in photographing the more delicate prominences upon the limb, it is better to exclude the light of the photosphere by covering its image by a circular disc. In the resulting picture the Sun consequently appears black, and the whole strikingly resembles a photograph of the eclipsed Sun.

In our brief study of the work of the spectroscope we have but touched upon those of its applications that have so far proved the most important, probably because it has been found possible to interpret them. It has been necessary to pass over a vast accumulation of its records, in some of which a meaning is indicated with less certainty, but which, in greater part, have utterly baffled rational conjecture. It cannot be imagined, however, for a moment that the story of the spectroscope is as a tale that is told. Year by year its record is accumulating, while year by year advance in other branches of physical science aids in the task of dealing effectively with it. The story of its work during the past fifty years is, however, alone a noble record of scientific achievement; and it is with feelings of highest interest and keenest expectation that the astronomer is now watching its continual development. But the path that we have followed with some care is one only of a number along which knowledge is advancing with no less success and promise of future triumph. At the close of the nineteenth century as never before does the music of Nature resound with a soul-inspiring harmony, and never in the past have the paths of science appeared so exquisitely attractive to her children.

Index.

Aberration of light, 50, 51.
Absorption, mechanism of, 177.
Algol, discovery of dark companion, 22, 213.
Alpha Centauri, distance of, 6.
Alpha Crucis, relation to Milky Way, 90.
Alpha Cygni, relation to Milky Way, 86.
Andrews, critical temperature of gases, 37.
Andromeda, nebula in, 16.
Ångstrom, explanation of dark lines in solar spectrum, 179; method of observing spectrum of Sun, 219; researches on chemistry of the Sun, 186.
Arcturus, motion of, 46, 70.

Barnard, observations of Mars, 120, 121, 129; photographs of the Milky Way, 66, 67.
Becquerel, photographs ultra-violet region of spectrum, 167.
Bessel, first detects parallax of a star, 5, 50.
Bifurcation of Milky Way, 60.
Boeddicker, drawing of the Milky Way as seen by the naked eye, 89, 131.
Break in the Milky Way, 62.
Brewster, discovers telluric lines in solar spectrum, 166.
Bunsen and Kirchhoff, researches in spectrum analysis, 180.

Calcium, incandescent vapour of, in faculæ, 235; variation of spectrum under different physical conditions, 190.
Carbon, discovery of vapour in atmosphere of Sun, 187.
Carbonic acid gas, suggested as a constituent of atmosphere of Mars, 142.
Clustering of stars in neighbourhood of Sun, 97.
Coal Sack, in Milky Way, 61, 76, 81, 83, 89.
Collisions between stars, 45, 48.
Colour, effect of motion of source upon, 29.
Colour, relation to length and frequency of ether waves, 173, 175.
Copernicus, his view regarding the nature of the stars, 2.
Corona, 221; drawings of, 131.

Daguerre, his discoveries in photography, 167.
Dark matter in space, possible existence of, 13, 32, 67.
Dark stars, 13, 32; possibility of detection, 22, 35.
Diffraction grating, 160.
Doppler, enunciation of principle regarding the effect of motion of source on generated waves, 29, 200.
Doppler's principle, application to the discovery of Algol's companion, 29.
Douglass, his observations of Mars, 120, 126.

Draper, photographs infra-red region of spectrum, 168; photographs the Moon, 167.

Earth, appearance of, as seen from a planet, 114.
Eclipse of 1868, 222.
Energy, conservation of, 45.
Ether of space, 172.

Faculæ, solar, 234, 235.
Fizeau, indicates correct application of Doppler's principle, 29, 206.
Foucault, observes absorption of D light by gases of electric arc, 168, 170.
Fraunhofer, his improvements in the spectroscope, 156; observes dark lines in spectra of stars, 157; observes dark lines in spectrum of Sun, 156.
Fraunhofer lines, first observed by Wollaston, 155; studied by Fraunhofer, 156.

Gamma Cygni, relation to Milky Way, 87.
Goodricke, suggests eclipse theory of Algol, 23.
Groombridge 1830, 46, 99.

Hale, photographs prominences and faculæ in uneclipsed Sun, 232.
Heliometer, 54.
Helium, discovered in clevite by Ramsay, 225; a constituent of some nebulæ, 200; in solar atmosphere and prominences, 225.
Henderson, detects parallax of alpha Centauri, 6.
Herschel, Sir John, his drawing of the Eta Argus nebula, 130; his views regarding the structure of the sidereal system, 80; observes the relation between the nebulæ and Milky Way, 71.
Herschel, Sir William, observes antipathy between nebulæ and stars, 71; observes relation of stars to the Milky Way, 68; views regarding the structure of the system of the stars, 73, 77; views regarding the Sun, 36.
Huggins, Sir William, demonstrates gaseous nature of certain nebulæ, 21; determines motion of stars in the line of sight, 30, 206; observes variations in the spectrum of calcium, 191; observes spectra of stars, 196; observes prominences in uneclipsed Sun, 227.
Huygens, enunciates wave theory of light, 172.
Hydrogen, in atmosphere of Sun and in solar prominences, 225; spectrum of, 186.

Janssen, observes spectra of prominences in uneclipsed Sun, 222.

Keeler, his observations of Mars, 129.
Kirchhoff, his researches on the chemistry of the Sun, 184; his researches in spectrum analysis, 180.

Lane's law regarding the variation of temperature of a cooling gas, 38–40.
Light, possible absorption of, in space, 79; refraction of, 147; wave theory of, 172–175.
Lockyer, Sir Norman, his researches in the chemistry of the Sun, 187; his dissociation hypothesis, 191; meteoritic hypothesis, 43 (note).
Lowell, his observations on Mars, 166.

Mars, 101; atmosphere of, 110, 113, 114; canals of, 102, 124, 132; climate of, 132, 142; clouds on, 122; dark belt surrounding polar

Index.

cap on, 121; distance of, 103; gravitation on, 103; gray-green "seas" on, 117, 120; limb-light on, 110, 112; "a miniature of the Earth", 102; mass of, 103; oppositions of, 108; orange continents on, 117; phases of, 104; polar caps of, 116; rotation of, 108; seasons on, 109; size of, 103; speculations on possible inhabitants, 127; telescopic appearance of, 108; vapour of water in atmosphere of, 115.

Milky Way, Barnard's photographs of, 66, 67; a collection of faint stars, 60; dark rifts in, 67, 81; a definite formation, 64, 82; distance of, 62, 92; early views regarding, 59; general appearance of, 59; lines of stars in, 67; nebulous matter in, 64, 68, 87; photographs of, 65; structure apparent in, 63; telescopic appearance of, 60, 64.

Moon, nature of motion round Earth, 25.

Nebulæ, constitution of, 197; demonstration of gaseous nature of, 19, 197; early conjectures as to nature of, 16; Herschel's views regarding, 16; regarded as external galaxies, 18; relation of, to Milky Way, 71; temperature of, 38.

Newton, Sir Isaac, his analysis of sunlight, 144.

Oppositions of a planet, 104.
Orion, nebula in, 13; star streams in, related to Milky Way, 88.
Oxygen, its apparent absence from the atmosphere of the Sun, 192.

Parallax, method of relative, 3, 52, 56; of a star, 5, 51, 52.

Photographs, of infra region of spectrum, 168; method of taking, of celestial objects, 66; of Milky Way, 66, 67; of spectrum, 167, 168, 187; of ultra-violet region of spectrum, 167.

Photography, discovery by Daguerre, 167; invention of gelatine plate, 212.

Photosphere, solar, 36.
Pickering, E. C., his discovery of spectroscopic double stars, 215.
Pickering, W. H., his observations of Mars, 121, 126.
Prism, action of, upon light, 149.
Proctor, maintains relation of lucid stars to Milky Way, 69, 86, 89; suggests possible structure of Milky Way, 83.
Prominences, solar, 228; connection with sun-spots, 229.

Radiation, mechanism of, 176.
Ramsay, discovers helium, 225.
Ranyard, maintains intimate relation between stars and Milky Way, 86.
Refraction, by atmosphere, 51; of light, 147.
Resonance, 177.
Reversal of spectral lines, first observed by Foucault, 168, 170; theory of, enunciated by Stokes, 171-179.
Rowland, his photographs of solar spectrum, 187.

Schiaparelli, discovers the canals of Mars, 124.
Selective absorption, 135, 138.
Simms, applies collimator to spectroscope, 157.
Sirius, brightness of, 8; distance of, 7; motion of, in line of sight, 206, 209; spectrum of, 207.
Sky, cause of appearance of, 110.

Spectra, conditions of purity of, 152, 157, 158; of flames, 163, 165; of solar prominences, 222; variation of, with different physical conditions, 188; of nebulæ, 198; of stars, 194; of Sun, 144.

Spectro-heliograph, 236.

Spectroscope, principle of, 153; prismatic, 158.

Stars, death stage of, 22; distances of, 3-8, 50; hypothesis of uniform distribution of, in space, 77, 78, 94, 96; magnitudes of, 92; motion of, 45; motion of, in the line of sight, 213; relation of, to Milky Way, 68, 71, 82, 86, 91, 92, 98; spectra of, method of obtaining, 194; spectroscopic double, 215; sun-like nature of, 2, 8.

Stewart, Balfour, states condition necessary for reversal of spectral lines, 182.

Stokes, Sir Gabriel, explains reversal of spectral lines, 171-179.

Struve, Wilhelm, his views on structure of sidereal system, 81.

Sun, death stage of, 11; life history of, 42; motion of, in space, 47; source of heat of, 11; telescopic appearance of, 35.

Sympathetic vibration, 177.

Telluric lines in solar spectrum, 166.

Tyndall, his researches on selective absorption, 137; illustrates the theory of sky formation, 110.

Universe, last catastrophe of, 49.

Vogel, demonstrates existence of dark companion of Algol, 24, 31; measures motions of stars in the line of sight, 212.

Waters, Sydney, his map of nebulæ in their relation to Milky Way, 72.

Waves, formation of, 176; of light, 172-175; of sound, 174, 200.

Wollaston, first observes dark lines in solar spectrum, 156; improves conditions for viewing spectra, 155.

Wright, of Durham, his theory regarding the structure of the stellar system, 73.

Young, C. A., observes reversal of calcium lines in faculæ, 235.

Young, Thomas, establishes the wave theory of light, 172.

*In Four Half-vols., cloth, price 12s. 6d. each, nett; or
Two Vols., cloth, 25s. each, nett.*

THE
NATURAL HISTORY OF PLANTS

THEIR FORMS, GROWTH
REPRODUCTION, AND DISTRIBUTION

FROM THE GERMAN OF
ANTON KERNER VON MARILAUN
Professor of Botany in the University of Vienna

BY
F. W. OLIVER, M.A., D.Sc.
Quain Professor of Botany in University College, London

WITH THE ASSISTANCE OF
MARIAN BUSK, B.Sc., AND MARY EWART, B.Sc.

*With about 1000 Original Woodcut Illustrations and
Sixteen Plates in Colours*

KERNER'S NATURAL HISTORY OF PLANTS, now for the first time presented to English readers, is one of the greatest works in Botany ever issued from the press. Its province is the whole realm of Plant Life, and its purpose, as conceived by the author, Professor Anton Kerner von Marilaun, of Vienna University, is to provide "a book not only for specialists and scholars, but also for the many".

OPINIONS OF THE PRESS.

"The best account of the vegetable kingdom for general readers which has yet been produced. . . . The translation is scientifically accurate, as well as entertaining and instructive. Lovers of nature will find every page of the book interesting, and the serious student of botany will derive great advantage from its perusal. The illustrations are beautiful, and what is more necessary, true to nature."—**Nature.**

"The first number exhibits a perspicacity of treatment and a simplicity of expression which we are wont to consider as rather uncharacteristic of the German intellect in its deeper workings. Indeed, the treatment is as popular as it is conceivable for it to be, and the paucity of technical terms a feature very highly to be commended."—**Journal of Horticulture.**

LONDON: BLACKIE & SON, LIMITED; GLASGOW AND DUBLIN

THE
Warwick Library of English Literature

In crown 8vo volumes, cloth, 3s. 6d. each.

General Editor—Professor C. H. HERFORD, LITT.D.

Each volume in the present series will deal with the development in English literature of *some special literary form*, which will be illustrated by a series of representative specimens, slightly annotated, and preceded by a critical analytical introduction.

English Pastorals. With an Introduction by E. K. CHAMBERS, B.A.

English Literary Criticism. With an Introduction by C. E. VAUGHAN, M.A., Professor of English Literature at University College, Cardiff.

English Essays. With an Introduction by J. H. LOBBAN, M.A., formerly Assistant Professor of English Literature in Aberdeen University.

English Lyric Poetry (1500-1700 A.D.). With an Introduction by FREDERIC IVES CARPENTER, M.A., Lecturer in English Literature at Chicago University.

English Masques. With an Introduction by H. A. EVANS, M.A., Balliol College, Oxford.

IN PREPARATION:

English Letter-Writers. With an Introduction by W. RALEIGH, M.A., Professor of English Literature at University College, Liverpool.

English Tales in Verse. With an Introduction by C. H. HERFORD, Litt.D., Professor of English Literature at University College, Aberystwyth; General Editor of the Series.

OPINIONS OF THE PRESS.

"The idea of such a series has much to recommend it, and it is well carried out in this comely and attractive volume ('English Pastorals'). In his introduction Mr. Chambers writes of the history and characteristics of the Pastoral with learning, insight, and sympathy."—**The Times.**

"It is refreshing to find an editor who can write with just appreciation and without exaggerated praise. Mr. Lobban (in 'English Essays') has accomplished his task with care and good judgment, the consequence is that he writes with a certain fulness, and thereby seldom fails to interest his reader."—**Journal of Education.**

"This excellent addition ('English Lyrics') to an excellent series deserves notice and commendation. The notes, literary and biographical, which are prefixed to the selections from the different authors, are notable for conciseness. The introduction is also a sound piece of criticism, tracing with great clearness the connection between the lyrical impulse and performance of particular periods and the national history."—**Spectator.**

LONDON: BLACKIE & SON, LIMITED, GLASGOW AND DUBLIN.

www.ingramcontent.com/pod-product-compliance
Lightning Source LLC
Chambersburg PA
CBHW021409230426
43666CB00006B/684